Planning and Building Vacation Homes

Ronald Derven
Ellen Rand
James Ritchie

Ideals Publishing Corp.
Milwaukee, Wisconsin

Table of Contents

Introduction 3

Finding a Place in the Country 4

Hiring a Builder, Drawing the Plans 10

What Type of Home? 14

Preliminary Steps 19

Building It Yourself 25

Building a Log Cabin 59

Doors, Windows, and Decks 74

Kits, Prefabs, and Packages 78

Energy Saving Tips 82

Space Savers and Maintenance 87

Protecting Your Vacation Home 91

Glossary 94

Index 96

ISBN 0-8249-6114-5

Copyright © 1982 by Ronald Derven, Ellen Rand, and James Ritchie

All rights reserved. This book or parts thereof may not be reproduced in any form without permision of the copyright owners. Printed and bound in the United States of America. Published simultaneously in Canada.

Published by Ideals Publishing Corporation
11315 Watertown Plank Road
Milwaukee, Wisconsin 53226

Editor, David Schansberg

Cover photo courtesy of Pan Abode

⌂ SUCCESSFUL
HOME IMPROVEMENT SERIES

Bathroom Planning and Remodeling
Kitchen Planning and Remodeling
Space Saving Shelves and Built-ins
Finishing Off Additional Rooms
Finding and Fixing the Older Home
Money Saving Home Repair Guide
Homeowner's Guide to Tools
Homeowner's Guide to Electrical Wiring
Homeowner's Guide to Plumbing
Homeowner's Guide to Roofing and Siding
Homeowner's Guide to Fireplaces
Home Plans for the '80s
Planning and Building Home Additions
Homeowner's Guide to Concrete and Masonry
Homeowner's Guide to Landscaping
Homeowner's Guide to Swimming Pools
Homeowner's Guide to Fastening Anything
Planning and Building Vacation Homes
Homeowner's Guide to Floors and Staircases
Home Appliance Repair Guide
Homeowner's Guide to Wood Refinishing
Children's Rooms and Play Areas
Wallcoverings: Paneling, Painting, and Papering
Money Saving Natural Energy Systems
How to Build Your Own Home

Introduction

To own a piece of secluded land and to build a vacation retreat on it is almost everyone's dream. But can that dream be turned into a reality at an affordable cost? This book is designed to give you basic guidelines in finding and buying property, and then planning and building an economical and totally enjoyable vacation home.

Land and building costs are high and will rise even more in the future. It has been estimated that land costs double in price every five years, while building material costs have often outdistanced the inflation rate. There are really no bargains to be had in buildable land located in desirable beach, mountain, woodland, or lake areas; nor are high building costs likely to decrease in the foreseeable future. However, by following guidelines set forth in this book, you will be able to transform your dream vacation home into a reality as economically as possible. It should be noted that if you decide to wait just another few years to start building, future construction and land costs may well outpace any dollars you have saved.

Budgets The key to building an affordable vacation home is to realistically scale your design, material preferences, and interior space requirements to suit your budget and basic needs. You may be surprised to find that your needs can be comfortably accommodated in a home of less than 500 square feet or that a parcel of land within walking distance of a lake can be less expensive, and perhaps have better soil drainage than land with beach frontage.

We point out, too, how to cut down on costs by doing much of the work yourself and by acting as your own contractor.

Alternatives Another way to achieve savings is to select one of the hundreds of manufactured home packages available: precut, prefabricated, modular, or mobile homes. You can assemble many of the shells of these home packages yourself with rudimentary building tools. Packaged homes can prove advantageous in areas where labor is in short supply or in cases where the shell must be put up quickly so that work can proceed indoors in inclement weather. You should determine what the shipping and interior finishing will cost before you choose a packaged home. Shipping can cost almost as much as the home depending on location.

Energy Savings Energy conservation strategies are stressed throughout the book. Alternative techniques are presented for heating, cooling, and insulating vacation homes for maximum energy efficiency. These basic techniques will also mean reduced operating and maintenance costs over the life of your home.

Planning Clearly, building a vacation home can present more puzzles than building a primary residence would present. Whereas in your primary residence you may not give a second thought to flushing the toilet or turning on a faucet in the kitchen, these basics must be carefully planned in a vacation home. Then, too, there is the question of dealing with local officials: health department inspectors, planning and zoning boards (in many cases), and building inspectors. One of the biggest quandaries is keeping track of progress on your home when, for the most part, you may be hundreds of miles away.

Despite all these potential headaches, the effort will be worth it. Whether for a single person or a family, a vacation home offers more than just a place to get away from it all. It is a place to entertain friends and share experiences with those close to you. It is also a good investment; you can rent it out for extra income, or you can swap it with other homeowners in the U.S. and abroad for a rent-free vacation.

We intend this book will give you a realistic view of the pleasures and pains involved, and aid you in overcoming difficulties and avoiding problems encountered during the various buying and building stages.

Finding a Place in the Country

The enjoyment of your leisure home can be greatly enhanced by recollection of all the excitement and headaches involved in creating it. Vacations and weekends spent building the home will allow you to gain greater appreciation of the home when it is completed. Make no mistake; building a leisure home is not for the faint of heart. It requires tremendous time and energy—first finding a suitable spot to build, and then drafting and executing the plan. Patience, flexibility, and a sense of humor when confronted with unexpected situations will help maintain a pleasant outlook that will allow you to enjoy the construction of the home.

Early Considerations

Some of the problems that can arise involving the land itself are detailed in this chapter. Problems arise, too, during construction, particularly if severe weather hampers building progress or if material shortages are experienced.

One of the toughest aspects of building a leisure home is taking measures to protect it. Unless you are building it yourself, you are usually too far from the site to oversee daily construction and to keep watch on materials.

Even if you spend all your vacation time and weekends at the site, it's difficult to really gauge how effectively subcontractors are working on your house. If you are a newcomer to the area in which you're building, it is particularly hard to determine whether or not you are paying market rates for labor and materials.

Once the house is finished, you must concern yourself with securing it during the week, or during the seasons when you won't be using it. The seclusion you enjoy makes your vacation home easy prey for burglars and vandals. The home must also be protected against natural invaders like termites and wild animals, as well as against harsh natural enemies like wind, rain, snow, and water. Moisture in both its dramatic and unobtrusive forms can prove to be your home's worst foe, creeping into basement or crawl space, attics, doors, windows, and rusting out furnaces and water heaters.

Economically, your home can be a problem, too. If at some point you find it impossible to carry the extra cost of owning a leisure home, you may have to think of selling it. Although it is a tangible investment that generally appreciates in value at a greater rate than inflation, it is not a liquid asset. That is, if you have to sell it immediately, chances are you won't get the best price for it.

All this is not to dissuade you from embarking on this major undertaking. Rather, this may alert you so you can avoid these problems.

Vacation homes are perhaps the least documented segment of the housing universe. Average figures on resale values, mortgage requirements, and preferred styles are elusive. Because there is such a tremendous diversity in the kinds of properties (beachfront to mountaintop) and in the kinds of homes built (precut geodesic domes to Swiss-style chalets), generalizations are not possible.

Where to Build

According to the latest figures available from the *Annual Housing Survey,* a joint publication of the Census Bureau and the Department of Housing and Urban Development, here is a breakdown, regionally, of the vacation home market.

	Vacant, Seasonal, and Migratory	% of Market
Northeast	591,000	38.5%
North Central	424,000	27.6%
South	341,000	22.3%
West	178,000	11.6%

Climate The design of your house and the materials used to build it will be influenced greatly by the climate in the area.

Sunshine—Coastal areas generally have more cloudiness than interior areas. Mountainous areas have less sunshine than interior areas, while desert areas have the most sunshine of all. During November, December, and January, the Pacific Northwest gets about 100 hours of sunshine, while the Great Lakes area gets about 80 to 100 hours. Southeastern California and southern and western Arizona get more than 240 hours of sunshine. In the summer, most of the U.S. gets about 300 hours of sunshine per month. Again, the greatest amount is in southeastern California and southern Arizona. The least mean total sunshine is experienced along the Washington and Oregon coasts: 1,800 to 2,200 hours annually.

Low in square footage, yet functional and dramatic, this home stresses indoor/outdoor living with a deck off the kitchen for dining. Windows are located and designed to *let in maximum amounts of natural light and heat; the interior chimney is another energy-saver. Photo courtesy of Champion Building Products*

Humidity—The amount of moisture in the air can either make you feel hot and uncomfortable in the summer, or cold and clammy in the winter. The preferred indoor relative humidity range is 40 to 60 percent.

Relative humidities are highest in coastal areas where winds blowing inland from water surfaces carry a lot of moisture. Humidity is lowest in deserts and the semiarid Southwest. While Tampa, Florida, may be suffering 88 percent humidity, Yuma, Arizona, would have perhaps 57 percent. The Northeast is generally more humid than the Southwest.

Wind—On-shore winds over coastal areas can make those areas milder and moister than winds coming from the interior of the continent. Prevailing winds on the Pacific Coast are from the northwest. Winds over the eastern two-thirds of the country are from the northwest or north during January and February, and generally from the south and southwest from May through August. Easterly winds of northeast trades prevail over the Florida peninsula except during December and January, while southerly winds prevail in the Texas-Oklahoma area from March through December.

Fog—Heavy fog, in which visibility is reduced to one-quarter mile or less, occurs on from 80 to 100 days annually in several areas, including: the northern part of the Pacific Coast; along the California coast; in the Olympic and Cascade ranges in the West; in the Appalachians; and along the New England coast.

Floods—The northern interior and northeast parts of the U.S. get floods most frequently during the spring months, while the Pacific states experience flooding more in the rainy winter months.

Snow—Mean annual snowfall at the higher elevations in the western mountains ranges from 100 to over 200 inches per year. The upper Great Lakes region and northwestern Lower Michigan experience 20 to 30 snow days per year, as does most of New York and New England and parts of Virginia, Maryland, and Pennsylvania.

Temperature Variations—Temperatures can vary widely over short distances, especially in mountainous areas. Variations are due to differences in altitude, slope of land, type of soil, vegetative cover, bodies of water, air drainage, and urban heat effects. An increase in altitude of 1,000 feet can cause a decrease of 3.3 degrees Fahrenheit in the average annual temperature of an area.

You can get more specific information on the climate of your chosen area, including precipitation records, humidity, temperatures, wind, fog, and other factors from the U.S. Department of Commerce, Records Section, Asheville, NC 28801. Climate books are published for each state.

Choosing the Site Try to locate property within several hours of your primary residence. This means you will spend less time and money driving back and forth, both at the choosing and using stages.

Future gasoline crises aside, it is difficult to enjoy your leisure time after a too-lengthy drive, facing the prospect of a similar drive back. During construction, particularly, you'll be traveling back and forth frequently enough to discourage a very long drive.

Once you know the type of property and area you'd be happiest in, start checking local sources for leads on available acreage. Officers of local lending institutions, classified ads in real estate sections, building-supply outlets, and town or county officials (town clerks, zoning, planning, building departments) can help.

Area real estate brokers are probably the most well-informed group about the market and current prices. A good broker can not only steer you in the direction of the property; he or she can help with negotiations for purchasing the property and can suggest methods and sources of financing. But remember that a broker acting as an agent for the seller has the responsibility of selling that property at the best price, quickly. Don't allow yourself to be rushed into a purchase. Don't sign any contract or purchase any property if there are still unanswered questions.

Some of the points to be covered include:
- Is there a year-round source of water? If a well must be dug, will you have the legal right to use that water?
- Is the ground conducive for building? If there is no sewer system, can the soil accommodate a septic system? Is drainage good, or will you have a problem with flooding after a rainstorm? Is the property on a floodplain?
- Can utilities be brought into the site easily? Will the cost of building an access road, bringing in electricity or gas, wipe out economies achieved in buying an out-of-the-way parcel?
- If the property is adjacent to public lands, are there any development plans afoot that would affect the site? Does the town or county plan on building near the site?
- If the property fronts on a country road, does the road provide direct access to the site or is there, perhaps, a ravine fronting on the road?
- What kinds of building restrictions are there? Is there minimum parcel zoning or minimum floor space requirements? Can you build a home for four-season use, or is seasonal use restricted?
- Are there restrictions on architectural style?
- Are any building moratoriums in effect? (This affects coastal properties in particular.) Is there any ban on additional gas or sewerage hookups?
- Must you submit some form of environmental impact statement to local or state agencies for approval?
- If other properties have to be crossed to get to the site, will you have the legal right to do so? Is such an easement obtainable?
- If others have to cross the site to get to their own properties, will you be disturbed? Can a house be set far enough from the road to prevent any disturbance?
- What are the taxes? Are there any existing liens on the property?
- Is the site within easy reach of police and firemen?
- Are recreational and cultural activities close by? Where are suppliers of basics like groceries and hardware?
- If the property includes a lake, pond, or stream, is the water free from contamination?

Try to visit the property under less-than-ideal conditions, such as during an off-season or in the rain.

Real estate professionals say that location is the most important factor in determining value. Even the professionals sometimes make mistakes in judgment of location and price when acquiring property, so you should understand at the outset that finding the right land may be a painstaking process.

Land is generally less expensive if you buy several acres, rather than buying one acre or one lot. In the same area, a one-acre parcel might sell for $7,000, but a six-acre parcel might sell for $27,000, which is the better value in the long run. Moreover, you generally get more value for your dollar if you buy acreage directly from a landowner rather than from a land development company.

In calculating the price of raw land, keep in mind that building a road, bringing in utilities, drilling a well, and laying out a septic system can add several thousand dollars to the initial cost. Acreage that already has utilities in, but costs more, might be a more desirable alternative. Also, be sure that your purchase depends on getting whatever zoning variances you may need, on getting satisfactory results from soil percolation tests or well samples, and on getting clear title to the property.

A title company, probably engaged by your attorney, must ascertain that the seller has the right to sell the property to you. It is also a good idea to have a survey of the property done. This will cost a few hundred dollars, but will save you far more than that in potential legal headaches should there be a question over boundaries at some point in the future. The survey will determine the precise measurements of the land parcel. That description should be included in the contract of sale. Do not sign anything until you have spoken to your attorney.

Buying the Land Purchasing the property can be done in a number of ways; most vacation area land sales are cash transactions, but you are certainly not restricted to paying cash for the property.

One popular method of financing is through the seller. In other words, you pay a substantial down payment on the property, perhaps 40 to 50 percent, and arrange with the seller to pay off the remainder to him over a specified period of time, at an interest rate that is mutually acceptable. Terms might include interest payments only for a certain period of time, with a balloon principal payment at the end of that period. Or, you could arrange a schedule of conventional principal-and-interest payments.

If the seller has a mortgage on the land, you can either buy the property subject to the mortgage or assume the loan yourself. In the former method, the seller would still be responsible for making mortgage payments; your down payment and subsequent monthly payments would reflect this. If you assume the mortgage, that means that you are responsible for the monthly payments, at the same terms called for in the seller's mortgage. You would have to work out a mutually acceptable sum to pay the seller, above the existing mortgage.

Some banks don't allow assumption of mortgages, mainly because prevailing interest rates are likely to be higher than they were when the original loans were made. So, make sure that the lender gives written approval to the seller to transfer the property to you.

There are two stages to go through when buying real estate: signing a contract of sale and closing, at which time title is transferred. The period of time between "going to contract" and closing varies, depending on how long you and your attorney think it will take to satisfy all conditions.

When you sign the contract, you will have to put down a certain amount of money as an earnest money deposit. This is applied to the total agreed-upon amount for the down payment or sale price which you will pay to the seller at the time of closing.

Title is transferred to you by means of a deed. There are several different types of deeds. Try to avoid a quitclaim deed. This merely passes on any title, interest, or claim the seller has in the property, without guaranteeing that such title, interest, or claim is really valid.

Financing You will probably have to finance a portion of your vacation home. There are a number of ways and a number of sources available, but the question really is how much of a monthly payment can you afford beyond your regular monthly budget?

Most manufacturers or dealers of precut or prefab homes can help you arrange financing. They can either offer a list of possible sources or help you with your presentation to the bank of your choice.

If you seek a conventional long-term loan for your house, either through a local savings and loan, savings bank or commercial bank, or through your home town bank, be aware that terms for second homes are considerably stiffer than they are for primary residences. For a second home, a mortgage lender may require a down payment of up to 50 percent of the purchase price. Interest rates are generally .5 to 1.5 percent higher than for a mortgage on a primary residence. Moreover, a mortgage on a primary home is usually for a period of 25 to 30 years, while a mortgage on a second home is more likely to be for a period of 15 to 20 years.

On a construction loan, the bank will charge a fee, usually a small percentage of the loan. Funds will be advanced after the bank has been assured by a building inspector's report that each stage has been properly completed.

If money conditions are tight in the area or in the country generally, bankers are likely to cut down on making loans for vacation homes. They are considered a luxury, not a necessity or a priority item. Regardless of money conditions, your mortgage lender will carefully scrutinize your house plans to make sure they conform to their standards, and will carefully scrutinize your financial picture to assure themselves that you will be able to meet the mortgage payments.

If you want a standard construction loan (a short-term loan), you must own your property outright, and you must submit exact plans and specifications to the bank for approval. The interest rate is higher for a construction loan than it would be for long-term financing, and you will probably be required to have a licensed contractor finish the job to meet the bank's requirements.

Other ways of financing a vacation home are: taking out a personal loan; borrowing on your life insurance; refinancing your existing mortgage, or taking out a second mortgage on your primary residence.

One lesser-known source of financing is through Federal Land Banks. There are twelve Federal Land Banks in the U.S., one in each of the twelve Farm Credit Districts. These banks make long-term loans secured by first liens on real estate in rural areas through more than 500 local Federal Land Bank Associations.

Though designed primarily to finance individual and corporate farmers and ranchers, the Federal Land Banks do provide long-term loans for individual rural residents, too. To qualify, you cannot have more than one loan outstanding on a rural home at any one time. Nor can the loans be made if you're planning to build a home in order to rent or resell it.

A rural area is defined as being essentially agricultural in character, but may also include towns or villages with populations of less than 2,500 people which are not adjacent to large population centers.

Loans can be made for moderate-priced dwellings that are conventionally built, or are modular in construction, or for mobile homes meeting certain standards. Loan terms are flexible, and there is no prepayment penalty.

Renting Out and Tax Consequences

At some point you will probably hear that your vacation home will pay for itself if you rent it out during periods when you do not use it yourself. This is difficult, if not impossible, although renting can certainly help defray a portion of the cost of carrying that house.

Before you incorporate renting as part of your

This sensitive vacation home design uses natural surroundings to best advantage without disturbing the setting. The angular design of this Idaho vacation home is emphasized with redwood siding installed on the diagonal. Photo courtesy of California Redwood Association

This is an example of one of the "packaged" homes available; its roof overhangs the deck and carport providing style and shady areas for the owners. Photo courtesy of Deck Homes Inc.

vacation-home strategy, ask yourself these questions:

- Will you feel comfortable knowing that strangers are in your house? Extra maintenance and repair will be necessary to keep the house in top condition.
- Do you want to rent the property out during periods when it is most desirable, or would you prefer to use the home yourself during those periods?
- Is your property really marketable during the seasons you wish to rent it out?
- Is the rent level you have determined as necessary to cover your carrying costs in line with the market?
- Can you handle all the details of renting out the house yourself (i.e., advertising through classified ads, announcements in local commercial and retail establishments, word of mouth) and effectively "qualify" prospective renters or should you retain an area broker or management firm to rent the property for you?
- Will your homeowners insurance policy cover the home and its contents if it is partially used as a rental property?

The U.S. tax laws have become considerably more stringent regarding vacation homes. Since the Tax Reform Act of 1976, deductions attributable to the rental of a vacation home require that certain standards be met, namely:

1. The property must be rented for more than 15 days during the year in order for income and expenses to be reportable.
2. Deductions that amount to more than the gross rental income from the property are barred in cases where the owner rents out the home for 15 days or more and uses it for 14 days or more, or more than 10 percent of the number of rental days (whichever is greater).

In other words, you cannot effectively build up deductions such as depreciation greater than the amount of rental income. You can, however, prorate interest and taxes according to rental use.

Since tax laws can change frequently, it would be wise to check with an accountant about how rental of your vacation home affects your tax return.

Swapping As an alternative to renting out your house, you can swap your home with others for a rent-free vacation. There are a number of home-swapping services which have literally thousands of home listings in the U.S. and abroad.

Their directories have pictures of homes available for exchange purposes plus a brief summary of the homes' features. If you find something intriguing in one of these directories, it makes good sense to write detailed inquiry letters to prospective swap-mates, including photos of your own home and references.

Hiring a Builder, Drawing the Plans

If you are planning to do most of the construction yourself or will act as your own general contractor, this section is not intended for you. If not, it will be crucial to find a contractor who will understand your needs.

Because you are not likely to be at the site to supervise the proceedings on a day-to-day basis, it is particularly important for you to have a contractor in whom you have substantial trust. There are always unexpected situations in the course of construction: weather conditions can alter schedules; material shortages can delay deliveries and raise costs; subcontractors can go out of business. It's important that the contractor keep you informed and that you can rely upon him.

Finding a good builder requires a little research. Good sources include: friends or homeowners in your chosen area; your architect; the builders' association of the state; building-supply dealers in the area; lending institutions in the area; or local real estate brokers.

Pinpoint a few potential candidates, and ask each to give you rough cost estimates for your home. You should be able to give these contractors a pretty clear idea of the type of house you're interested in, the size, number of bedrooms and bathrooms, and types of materials you want to use. Remember that an estimate is not a commitment that your house can be built for that amount.

Be sure to check the builders' reputations with past customers. The builders should be willing to provide names. Ask if the work was accomplished within the proposed time frame, if there was sufficient attention to craftsmanship and detail, and whether the builder was easily accessible by phone. The local chamber of commerce and Better Business Bureau might be of assistance at this stage, too.

You should get at least three written bids for a home. These bids should detail your specifications: the scope of the work to be done, materials and supplies required, and complete costs. Make requests to each bidder the same, for comparative purposes. Set a time limit for receiving written bids. Three weeks to a month would be realistic. Do not assume that the lowest bid means the best deal for you. Quality is equally important.

Once you have selected a builder, a written agreement should be drawn up detailing the work involved and the payment schedule. Provision should also be made for an escrow account; that is, you will hold a certain portion of the final payment aside in case the work has not been completed to your satisfaction. Payment would be made when the work is completed to your satisfaction.

Payment schedules usually call for partial payment upon completion of a certain amount of work and other payments in stages as other phases of work are completed. Check over the final bills with a keen eye.

Acting as Your Own Contractor

If you are acting as your own contractor, setting up a payment schedule will be somewhat different. For carpenters who are going to be spending a long time on the job, it might make sense to establish credit at the local lumberyard or building-supply outlet. This way, the carpenter can order the materials and charge them to your account. This will give you an accurate record of the quantity and price of materials ordered. You would then have to pay the carpenters for their labor, calculated on an hourly basis.

Before work proceeds, come to an agreement about the timing of payments: whether it be weekly, twice a month, once a month, once every two months, or whatever is mutually agreeable.

Other tradesmen, such as plumbers, electricians, and masons, won't be spending as much time on your job as the carpenters. They will submit bills to you upon completion of their work, calculated on their hourly rate plus the cost of materials. Often these tradesmen will want to work for a set contract price which includes time, materials, profit, and overhead.

A rule of thumb is that labor will comprise about half the cost of building your house, which is one reason why it makes sense to do much of the work yourself, if you have the time and inclination.

Another rule of thumb is that building a home will cost approximately twice as much as the cost of the land.

How to Read Building Plans

Whether you are building the house yourself or having a contractor do it for you, understanding how to read building plans will give you the information on what must be done.

Vacation homes should be designed for minimum maintenance, as is this California beach house. Redwood siding need not be painted; vegetation around house is left in *its natural state. These features help keep maintenance at a minimum. Photo courtesy of the California Redwood Association*

For the typical single-family house, plans are broken down as follows: plot or site plan; foundation plan; floor plans; elevation plans; and mechanical plans.

Plot or site plan This shows the contours, boundaries, roads, utilities, significant physical features, large trees, and other structures on the site. Usually the surveyor will draw up the plan, which is then submitted to the architect. The architect in turn will design the structure and fit it on the survey. This plan takes zoning setbacks into account. Because existing elevations are also shown as contour lines, it allows you and the excavator to see how much earth must be moved.

Foundation plan This plan shows the main foundation of the house, as well as locations of piers. It also details the type of footing and concrete

block wall to be constructed. This plan will detail any waste or vent pipe which will go from the house to the outside. All dimensions are shown.

Floor plans This can be a single sheet of paper or many sheets of paper, depending on the size of the house. A small cabin, for example, would only take one sheet, whereas a three-story structure generally takes three or more sheets. Included in the floor plans are dimensions, location of windows and doors, location of walls and partitions, character of materials to be used, and more.

Framing plans These show the dimensions and arrangement of structural members. Often the floor framing plan is superimposed on the foundation plan to show sill plate, joist size and location, and bridging. The wall framing plan is quite similar in what it shows. That plan details the vertical plane

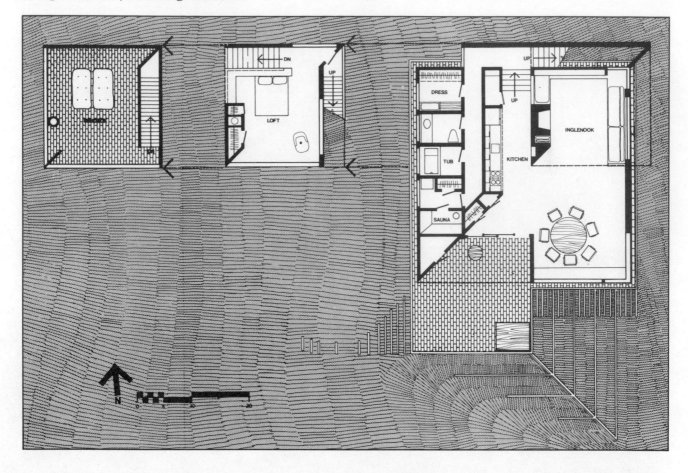

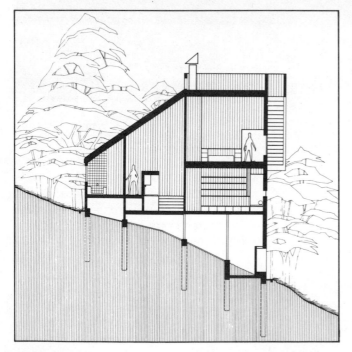

Architects' drawings are designed to show what a house will look like. Illustrative tools are elevations, floor plans, and contour drawings. Shown here are plans and the completed house, designed by architects McCue, Boone and Tomsick. The house takes advantage of a steep site and natural breezes by means of its unusual spiral-within-a-square plan. Photo courtesy of Jeremiah O. Bragstad

of the house, whereas the floor plan details the horizontal plane. Typically shown are studs, bracing, sill, plates, posts, and other structural members. The roof framing plan gives similar information regarding that structure.

Utility plan This plan details where utilities enter the house and location of electrical outlets and fixtures.

Mechanical plan If the house is large enough, the mechanical plan will detail where heating ducts are to be placed and the location of the furnace, hot-water heater, and other mechanical equipment.

Elevations These will probably be of greatest interest to you. Elevations show you in pictures what the house will look like. In fact, the intent of these architectural drawings is to communicate the scope of the house in illustrations.

Cross sections Often plans will detail various cross sections—cuts of a vertical plan. On the foundation plan, there may be a cross section of the concrete footing and foundation wall. Cross-section plans can detail other elements of the house as well.

Details Often the above-mentioned drawings can not accurately communicate a particular detail of the house. Therefore, the architect draws a very specific sketch of what he wants. This might be a particular wall section, interior stairs, or a deck railing.

House Construction Sequence

Homes do not appear magically. They are built according to a logical sequence of building tasks. A conventionally built house is constructed in this order:

1. building the foundation;
2. framing the walls and roof;
3. sheathing of the walls and roof;
4. applying the roofing;
5. installing exterior doors and windows;
6. applying exterior siding.

Each of these steps has to be completed satisfactorily before work can proceed on the next step.

After the shell is complete, work can proceed inside the structure despite poor weather. This will cut down on construction time. To finish the interior, the following order is followed:

1. partitions framed;
2. rough wiring, plumbing, and bathtub installed;
3. interior door frames put in;
4. wall and ceiling material applied;
5. electrical and plumbing work finished;
6. kitchen cabinets installed;
7. interior and exterior painted.

Several of the latter steps can take place at the same time.

The time required for all these steps will depend on several factors, some of which you can control and many of which you can not. Winter can be a good time to schedule construction, for example, because contractors and workers are most likely to be in a slack season, and therefore, are more readily available for you. On the other hand, severe storms can block access to your site, freeze equipment, and cause delays in deliveries of materials.

The number of workers on a job will also affect the time involved. If you and your brother-in-law are framing the house, it will probably take longer than it would for a builder's crew to do the same. If you are using tradesmen, such as plumbers, electricians, or masons, a strike can bring your project to a halt.

Remember, if you buy a precut or prefabricated home, it may take just a matter of days for the shell to be completed, but interior work can take as long as for any conventional home.

What Type of Home?

The style and size vacation home you select will depend on a number of factors. Among these are:

- Geographical location and climate considerations.
- Availability of materials in your area of the country.
- Projected use of the home: one season, two seasons, or all-year long.
- Style of life and individual preferences.
- Budget.

This chapter will give you an overview of what types of homes are available, what types of materials are suitable for different environments, and what types of construction are most economical. But first, let's start by figuring out your space and design needs.

Space and Design Needs

Your space requirements will depend, above all, on the number of people using the home and the activities for which you will use it. Some people just need a small home in the country; others enjoy entertaining lavishly. Honestly evaluate your own preferences before planning your vacation home.

If you are single or if you have no children, you can build a functional plan of about 500 square feet on one level that includes a bedroom. Smaller cabins are possible also.

If your tastes are not that spartan or if you plan to entertain or accommodate several guests, you might want to consider a two-level plan that includes about 600 square feet on the first level and about 200 square feet of storage/sleeping/den space on the second level.

Loft sleeping spaces are particularly dramatic. You get a wonderful sense of spaciousness, looking down to the living area, and a connection to the outdoors not possible in an enclosed bedroom. These types of sleeping areas are not ideal, however, if there are children around; the openness of the space means that noises will carry throughout the house.

If you do have children, figure on about an extra 80 to 100 square feet for every extra bedroom. It may not be necessary to add an additional bedroom for each child; if they are the same sex, they can probably share sleeping accommodations. An extra bedroom that goes largely unused is very costly in a leisure home.

One guiding principle stands out when trying to determine your space requirements: less is better. You should be more concerned about creating a functional, pleasant environment than building a testament to creative architectural thought.

Appliances such as dishwashers, garbage disposals, and trash compactors, for example, add an extra expense and drain energy in a second home. In bathrooms, curb your appetite for luxury. Plan instead on arranging the room so that all of the fixtures are on one wall, since this is most economical. Minimum space for one-person bathrooms would be 5' x 7'2". A family's needs could be accommodated in a bathroom of 5' x 8', 5'6" x 7'4", or 5'6" x 7'8".

Although you are urged to "think small and economical" when it comes to most building and design questions, don't stint in the quality of the materials of the house: electric wiring, plumbing fixtures and fittings, furnaces, pumps, or lumber.

Climate

When it comes to determining the heating and air conditioning requirements for your house, again keep in mind its potential use and the climate. Do you really need heat in a home that you won't be using in winter or air conditioning in a home shaded by trees or cooled by sea breezes? A house intended for year-round use, of course, will need a more elaborate system than would a one- or two-season cottage.

Let's take a look at how climate will affect the design and materials used. As examples, we'll focus on homes for three popular, and very diverse, areas: beach, ski country, and desert.

Beach Homes Beach homes must be built to withstand the considerable beating they will take from the wind, salt, intense sun, and ocean spray.

More and more municipalities are preventing new construction from being built directly on the oceanfront, for many good reasons. The beach environment is particularly unstable; tides naturally erode beachfront, while dunes naturally shift and get overgrown. Jetties impede this natural process by helping to keep an existing beach in place.

Severe rainstorms, ice storms, and hurricanes have a devastating effect on beachfront properties; many stunning and expensive homes have literally been washed away or crushed. Homeowners' insurance policies or in some cases Federal funds for disaster areas can help people rebuild their properties, but there are few experiences more heart-wrenching than finding your home destroyed.

A beach house, then, should be set high enough on piers to allow high tides to come in without disturbing it or set back far enough to allow a wonderful view over the dunes, protected from the elements.

To minimize the effects of salty air, be sure that the home uses metal supports, metal fasteners, and nails that have been galvanized or treated for corrosion and rust-resistance. For exteriors, if wood siding is used, it should be stained, not painted. The paint will only peel or chip. Stained siding weathers far better and improves with aging. Concrete and stucco are also good materials to use in this type of environment.

Beaches can be very windy. Decks should therefore be situated so that they are not directly in the path of prevailing winds. Awnings, vents, louvers, shutters, and outdoor furniture should be secure enough to withstand strong winds.

To prevent the sun from harming furniture and fabrics, use tinted glass or interior shades to block hot rays. The home should be situated on the site so that the bulk of the window areas will not be facing sunlight directly. Although it may seem strange to consider heating a home that gets most of its use in the warmth of summer, remember that coastal areas can be damp. A fireplace or heater can help dry out the atmosphere inside.

Ski Country If the sea, salt, sun, and wind are the bane of a beach home's existence, then heavy snow is the equivalent for a home in ski country.

Unlike beach homes, homes in ski country should be situated so that much of their roof area faces the sun in winter. This will make the interior warmer during those cold weekends. Situating the house into the wind also makes sense because there will be less likelihood of uneven snow buildup during snowstorms. Heavy snow swept to one side of the roof can cause structural problems due to too much weight. This is one reason A-frames are so popular in ski resort areas: their steep sides shed snow more easily than more conventional roof designs. Other roof designs that have been used in ski cottages are flat roofs (good for supporting heavy snow loads, but not so good in preventing leakage and providing optimum insulation) and no-eave roofs where outside walls are canted and indoor heat warms the roof edge to help melt the snow.

Snow buildup on the sides of the home (from heavy storms or from falling from the roof) can also be a problem. Snow can cover the main access to the home, so having another door at the second level or building the home 10 feet off the ground should be considered.

Access to the ski house after a snowstorm can be a problem, too. Garages located off the nearest road that is frequently plowed are an alternative.

Frozen pipes can be disastrous in a ski home. To prevent this, water pipes should be located below the frost line, allowing the snow to act as an insulator. Plumbing vents should be high enough so that the snow will not cover them even after a big storm. Pipes should be placed within interior walls or enclosed.

Like beach houses, ski houses which have wood exteriors should not be painted. The wood should be stained and allowed to weather naturally.

Desert Homes In the hot, dry climate of the Southwest, staying cool is one of the prime requisites in choosing building design and materials. Materials that will retain the evening's coolness and inhibit heat buildup during the day are concrete, stucco, and masonry. Desert homes should be situated on their sites to limit the expanse of windows facing west, because that side of the house will become uncomfortably hot during the day. Dramatic overhangs, balconies, and porches can act as sunshades which will keep you cooler during the day. Screened sliding doors can be used to separate a living area from the outside and, left open, can help circulate breezes.

Proper ventilation is essential in these types of homes; floor level vents can help circulate air in otherwise stuffy rooms. Vents should be located at a high point along the roof, since hot air rises. Ceiling fans can also be effective in circulating air.

Economical Designs and Materials

Even within the constraints of climate and personal style preferences, there is considerable latitude in selecting the design and building materials for a vacation home. These choices are most economical:

- Two-story homes are generally less expensive to build per square foot than single-story homes, mainly because less roof and foundation areas are needed to cover a like amount of living area. If you are building the home yourself, however, two-story homes offer far more accident possibilities for you and your helpers.
- Rectangular floor plans cost less per square foot to build than irregular floor plans, including L-shaped or U-shaped plans.
- Simple gable roofs are most economical. Ridges and valleys increase the roof's cost. Flat and shed roofs are less expensive but have poor drainage and high maintenance cost.
- Basements can provide extra living space if

they are dry, well ventilated, and well lighted; the same holds true for attics.

- Slab-on-grade construction is cheaper than crawl-space construction. But locating utilities equipment in the main level in the slab-on-grade home will take away precious living space.
- Walls made of materials that form both exterior and interior wall surface, like concrete block walls, are more economical than composite walls.
- Hardwood floors are cost-effective over a period of time, despite their high initial cost.
- Thick tile flooring costs more than the thinner varieties, but lasts longer. It may be more economical in high-use areas. Vinyl asbestos and sheet vinyl floor coverings can also be used.
- Plywood is the least expensive siding available and can be stained. Cedar shingles, cedar clapboard, aluminum and vinyl siding and redwood are progressively higher-priced. Check local lumber mills if you are in the Southeast, Northeast, or Northwest for availability of lumber. It may be less expensive than that available at local lumberyards.
- Asphalt shingles make inexpensive roofing.
- For window frames, aluminum is less expensive than wood, although wood may be more esthetically pleasing in your home design.
- Consider factory-hung doors for rooms that do not need extra soundproofing. Hollow-core metal doors are least expensive. Solid-core doors can be used for bathrooms.
- Leaving framing materials like roof rafters exposed gives the home a rustic look and is more economical.
- Fiberglass baths and shower stalls are less ex-

Homes in ski country must be designed to withstand heavy snow loads, without allowing wind to drift snow unevenly. Flat roofs are good for handling heavy loads. They may have problems with insulation and leaking, however. Photo courtesy of Libbey-Owens-Ford Company

pensive than standard units. Check your local building code to make sure they are allowed.

Materials like redwood, marble, slate, masonry, and stone can be beautiful in a vacation home. They can also be expensive and require craftsmanship for installation.

Home Styles

Once you have determined your space needs and budget, you can consider what kind of configuration most appeals to you to provide that space. A myriad of design possibilities exist: from the simplest one-room cabin of under 200 square feet which you can build yourself to A-frames, traditional Cape Cods, ultracontemporary wood or masonry fantasies, to log cabins, geodesic domes, and newly built or recycled barns.

Cabins If you are planning to spend much of your leisure time by yourself or with one person or outdoors, the basic one-room cabin may be your most economical and desirable choice.

In all home designs, good traffic patterns are essential. In the one-room cabin, this is even more crucial. A good rectangular design would have essential kitchen appliances, including a small sink, stove, and refrigerator, along one wall. A fireplace could be placed in a corner. Space stretchers such as a loft that serve equally well as sitting or sleeping spaces will make the interior that much more workable. A patio off the entrance will complement the indoor/outdoor style of life, while skylights and clerestory windows will bring in natural light and a sense of spaciousness. Another economical touch, which also enhances the rustic feeling, is leaving interior wood studding and rafters exposed.

Barns Several years ago, buying old barns and recycling them for use as primary or second homes, became tremendously popular. The appeal of barns stemmed from their simple, sturdy construction and vast interior spaces filled with stunning layout possibilities. Moreover, many barns were available on large tracts of land that had outlived their use as farm properties.

It is getting increasingly more difficult to find old barns to renovate. It is also harder to locate sources for old barn materials: shingles, hand-hewn beams, end-grain wood blocks, old clapboards, and barn boards. Rather than seek out an old barn, many people have subsequently chosen instant barns, either stick-built or packaged, to give them a taste of rustic life.

Construction costs are comparable to, if not higher than, other housing configurations. Key elements are lofts, exposed-beam ceilings, cross bracing, and rough-textured wood exteriors and interiors. They can be built as small as 1,200 square feet in size.

Wind and salt air can cause damage to beach houses. Among solutions are plexiglass protection for deck area; redwood siding that can weather and needs no painting. Plywood can be stained too. Photo courtesy of the California Redwood Association

Geodesic Domes Geodesic domes offer an interesting and unusual design alternative for vacation homes. Although lenders and building inspectors are somewhat skeptical about dome-shaped homes, in many parts of the country they have proven to be a sturdy and satisfying form of shelter.

Essentially, a geodesic dome uses triangular space frames to form a spherical structure. Although the home itself appears round, it uses no curved surfaces in construction. The frames are usually fabricated in a factory and can be bolted together at the building site.

Although dome homes seem to herald a twenty-first century vision of housing with their unusual shapes and factory fabrication, in reality much of the construction is quite conventional. In roofing, for example, normal building products such as red cedar shingles or redwood shakes or asphalt composition shingles can be applied. For insulation, materials applicable in any wood frame building can be used: glass wool blankets, rock wool, urea-formaldehyde, styrofoam or urethane. The same

In hot climates, keeping cool can be helped by stressing indoor/outdoor living. As shown, the breakfast area in this tropical vacation home, which runs along one side of a glass-walled living room, is shady and sheltered. Photo courtesy of Champion Building Products

holds true for plumbing, electrical wiring, kitchens, and interior wall surfacing.

Some heating systems that have been used in dome homes include baseboard systems, radiant heat in the slab, prefabricated fireplaces, panel-ray heaters, floor furnaces, forced air systems, and even solar systems.

Dome shells can be put together by four or five workers over a weekend, and construction tools needed are rudimentary; no cranes, hoists, or gin poles are required.

There are several advantages to the dome shape for interior design, and if it suits your style of life, this type of house could complement a wooded, hilltop, beach, or mountain environment. There are no columns, beams, trusses, or girders to obstruct the interiors, which means great flexibility in interior layouts. Interior partitions need not extend all the way to the dome surface and can be fabricated so that they can be moved around or removed totally.

Of course, there are disadvantages to dome homes. Some dome dwellers have found furniture arrangement tricky, while others bemoan the lack of vertical walls. You must make absolutely certain that a chosen package will meet the requirements of a local community before you build.

Custom Homes Custom-designed and built vacation homes are fewer in number than modest cabins, cottages, A-frames, and other home styles built conventionally or from packages. Unquestionably, a home that is designed specifically to suit your needs and habits can not only be visually outstanding but can also provide the kind of built-in amenities you have always dreamed about.

Custom homes, of course, are the most expensive to build. Any time you deviate from standard sizes of materials and fixtures, or require many cuts in wood framing members to accommodate unusual layouts, or choose materials like quarry tile for flooring or redwood for siding and decking, construction costs escalate.

If you want a home custom designed and built to suit your individual tastes, you should find an architect with whom you feel compatible. The architect will spend time talking with you and your family to get a feeling for the way you live and what you're looking for in a vacation home. He or she will also want to become very familiar with your site.

The architect can be very helpful in locating a mortgage lending source, a builder, and material supply outlets you may not be familiar with yourself. He or she can also watch construction progress on your house to make sure that it is being built to specifications.

Home Packages There is a home package for every popular housing style: simple cabins, A-Frames, Cape Cods, multilevel homes, plus unusual styles like geodesic domes and hexagons. Many packages can be customized, added onto, or put together like building blocks to create a unique look.

Here are a few key questions to ask any package dealer or manufacturer.

- How complete is the package? What materials are included in the package? Some manufacturers' home packages only include exterior walls and the roof truss system, while excluding interior walls and roof decking.
- What is the quality of the materials used in the precut home? Are these materials suitable to the climate in your chosen area?
- What is the estimated delivery time once the order has been placed?
- How far must the package be shipped? How much does this add to the cost of the home?
- How long will it take to build the house? How easy is it to erect: does it require much technical construction knowledge? How much labor is actually involved?
- What services are available prior to and after the sale? Is help available in obtaining financing and in supervising construction during the erection of the home?
- How long has the company been in business? What kind of reputation does it have with past buyers?
- How tightly constructed is the home when finished? Is it energy efficient? Can it be expanded or changed to accommodate alternate energy sources?

Stock Plans If you do not want to go to the expense of hiring an architect and having a home custom-built but do not want to restrict yourself to manufactured homes either, there is another alternative: buying stock home plans and having a builder cost it out and construct it for you.

Several firms specialize in providing home plans. These include blueprints, materials lists, and specification outlines. To select a home that seems most suitable, go through the suppliers' catalogs, which can be ordered for several dollars apiece. These catalogs usually provide floor plans, artists' renderings, and a brief description of the house. There are literally hundreds of plans available for vacation homes: chalets, A-frames, round houses, octagonal houses, traditional cottages, log cabins, and multilevels.

The plans themselves are somewhat expensive. The benefit of using stock plans for small, simple homes which you may want to build yourself is that the plans are already worked out for you. This will help you with your banker, your subcontractor, and your building-mates.

Preliminary Steps

Handling Site Improvements

Before you purchase any tract of land be certain that basic needs such as water, sewage disposal, and electricity can be reasonably accommodated. And, just as this type of verification represents the first step toward your land purchase, bringing in those utilities represents the first step in home construction.

Road Building How accessible is your property? Naturally, one of the prime considerations in building a vacation home is "to get away from it all"; you do not want to be so near a major thoroughfare that the noise and exhaust will disturb your serenity. One of the possible implications of buying property far from the beaten path, however, is that you will have to build a road to get to your homesite.

Before you do that, make sure that you have the legal right to cross other property to get to your own. Chances are, the party selling the property to you already has that right. But if it is just a verbal agreement between neighbors, you would be well advised to have it spelled out in writing, preferably in the contract of sale.

If a road already exists, you should set the easement right at 60 feet in width. That way, if a new road has to be built to replace the existing one, you will still have some protection.

If a road does not exist, that will be the first building priority. Without it, no one will be able to deliver building materials. Costs for building a road depend on the length of the route, the type of materials used, and the condition of the terrain. Contacting a road contractor to get an estimate of what it would cost to build the road would be wise before you buy the property.

A minimum dirt road, over flat, clear terrain, would cost more than $2 a running foot to build. A blacktop road under the same conditions would be about three times the cost of a dirt road. And, if shrubbery, trees, or boulders have to be removed, or considerable grading work has to be done, the cost is higher still.

In the course of building a road or driveway, particularly if you are cutting into hilly terrain, it may be necessary to hold the earth that has been moved by means of a retaining wall. Cinder blocks represent the most economical method of creating a wall, although they may not be as esthetically pleasing as, say, railroad ties. A retaining wall of railroad

ties can cost several thousand dollars or more, depending on the length of the wall and the number of ties needed.

Bringing in Utilities Your property must have a year-round source of water that is sufficient to meet your needs. The minimum requirement for home use is about fifty gallons per day per person.

There are a number of sources of water supply: some public utility companies provide water, while others have public water districts or private water companies you can hook into.

If your property has a lake, pond, or stream, these may also prove to be sources of water for your home. To find out whether or not they are safe, check your county health department, the local Farm Advisor, or the U.S. Geological Survey.

You might also want to check with the Water Resources Department in your state to find out whether or not there are any plans for building dams on rivers nearby, which might affect your property. The Fish and Game Code in your state will let you know how your stream is classified. Violation of classifications can result in heavy penalties.

If the water on your property is usable, you'll need a shallow pump, piping, and a foot valve that fits onto the end of the line, to tap into it. This kind of system does have its drawbacks. It is difficult to design the system to ensure that it will work in the winter, when the water might freeze. Problems arise, too, during long dry spells. Contamination can occur, too, if a neighboring home has sewage difficulties.

If you are not able to tap into a water supply through a utility company or nearby lake, pond, or stream, you will have to have a well dug on your property.

About 90 percent of all water beneath the ground's surface occurs in the top 200 feet. The average depth of all domestic water wells in the U.S. is less than 50 feet. Again, the state water resources department and local health department will be able to give you a pretty good idea of how deep you will have to dig before you reach water.

Water is found in three types of formations beneath the ground: in layers of sand, in layers of gravel, and in porous rock or in cracks in rock.

Finding a possible source of water on an individual tract of land is far from an exact science. The usual procedure for finding a good water vein is to have a well dug. One guideline to keep in mind, of course, is that the well will have to be located at least 50 feet away from septic fields, for health's sake.

When a well driller sets up his rig and starts digging in a possible site, there is no guarantee that he will actually find a water vein on the first try or subsequent tries. You will have to pay for the drilling regardless of the time it takes.

Besides having the well drilled, you'll need to have a pump installed and a water storage tank put into the house.

Unless you have a real taste for rudimentary living, you will also have to concern yourself with bringing in electricity to your site. Generally, 100 amp service should be sufficient.

The cost for doing this will depend on how far your site is from the nearest utility pole. Poles should be placed about every 200 feet across the property, and the cost of each pole and wire varies from $100 to $500 installed. If you believe that above-ground poles and wires are unsightly, check with the utility company about installing underground wiring. It costs more, but esthetically it en-

The average depth of all domestic water wells in the U.S. is 50 feet or less beneath the ground's surface. The way to tap a water vein is by drilling wells. Photo courtesy of Deeprock Mfg. Company

hances your property and will ultimately prove to be a plus if you ever decide to sell your home.

Handling Sewage Disposal

If your vacation home can not be connected with a public sewage disposal system, then you will need to install a septic tank sewage disposal system.

Not all types of property can accommodate such systems, however, so before you buy the property, find out what the land's absorption capacity is, and what type of absorption field will be required to handle your needs. Your building department will give you requirements for this.

Soil Percolation Tests Soil is tested for absorption capacity by percolation tests. These can be conducted by local health departments, civil engineers, or licensed firms in the business. The tests involve boring six or more holes of 4 to 12 inches in diameter and as deep as the proposed trenches or seepage bed, throughout the proposed absorption field. Sand and gravel are placed on the bottom of these holes.

At least 12 inches of water is poured into each hole, with water added to keep the water level 12 inches above the gravel for at least four hours, preferably overnight. The percolation rate is the drop in water level over a 30-minute period, multiplied by two.

In sandy soil, water seeps more rapidly, and the test is run over a shorter period, with the rate calculated during the final 10 minutes of a one-hour period.

Septic Systems A septic tank disposal system works this way: the effluent from the septic tank is

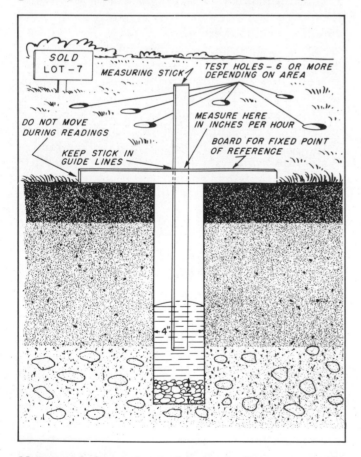

Many test holes, each 4 to 12 inches in diameter, must be dug for percolation tests. Tests indicate absorption capacity of soil, important in determining the size of the absorption field. Drawing courtesy of Soil Conservation Service, U.S. Dept. of Agriculture

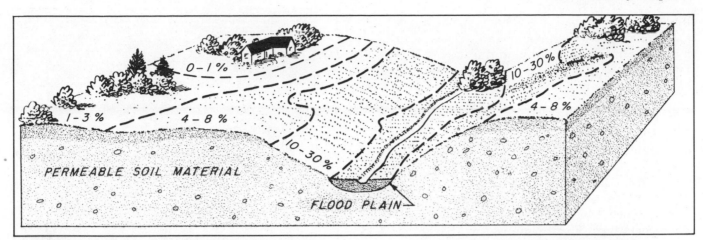

Deep, permeable soils are good for septic tank disposal systems, but slopes of more than 15 percent will present a problem. Floodplains are out as possibilities for absorption fields. Drawing courtesy of Soil Conservation Service, U.S. Dept. of Agriculture

carried through drain tile to points in the property where it is absorbed and filtered by surrounding soil. The drain tile is laid in trenches, which are then covered with soil.

The rate at which effluent moves into and through the soil is a major factor in determining how well a septic system will work. Soil permeability should be moderate to rapid, and your percolation rate should be at least 1 inch per hour. But there are other influences, too: ground-water level, soil depth, underlying material, slope, and proximity to streams or lakes.

At least 4 feet of soil material between the bottom of the trenches or seepage bed and any rock formation is needed for absorption, filtration, and purification of effluent. Deep, permeable soils are most desirable.

Floodplains are not suitable for absorption fields,

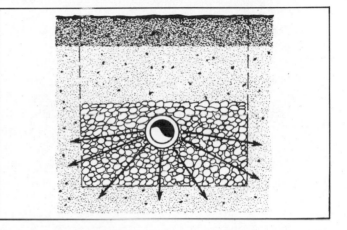

nor are slopes that have a grade of more than 15 percent. On more gently sloping lands, trenches can be dug on the contour so that effluent flows slowly through tile or pipe and disperses properly over the absorption field.

Your local public health department will be able to tell you what kind of standards it has set for absorption field sizes. These are generally calcu-

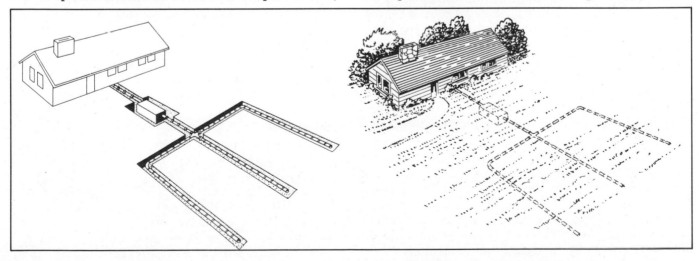

Effluent from the septic tank is absorbed and filtered by the soil in the field after being carried through drain tile

laid in trenches. Drawing courtesy of Soil Conservation Service, U.S. Dept. of Agriculture

lated by the number of bedrooms in the house.

For example, let us assume that you want to build a two-bedroom house on land that has a soil percolation rate of 2 inches per hour. According to the U.S. Department of Health, Education and Welfare, the area required for the absorption field is 250 square feet for each bedroom on property that has a soil percolation rate of 2 inches per hour. Hence, the absorption area required would be 500 square feet.

To determine the length of trench and tile or pipe required for this house, divide the total area by the trench width, 2 feet. Trenches should not be longer than 100 feet, nor should they be spaced closer than 6 to 8 feet apart. For this house, an absorption field layout might comprise four trenches of about 62 feet long or three trenches, each about 84 feet long.

To get the total area in square feet to be occupied by the absorption field, multiply the total trench length by the distance between trench and center lines. You will need a larger absorption field if the soil has a slower absorption rate.

Do not plan on laying an absorption field within 50 feet of a stream or lake, or your source of water can be easily contaminated.

Soil Surveys A soil survey can tell you the type of soil your property contains. These surveys can point out what the soil's limitations are and how it might be expected to perform as host to an absorption field.

Soil surveys of countries or other areas are published by the Soil Conservation Service of the U.S. Department of Agriculture. In fact, a booklet entitled "List of Published Soil Surveys" available from the department indicates that soil surveys are published for all 50 states, plus Washington, D.C. and Puerto Rico. The surveys contain maps as well as general information about the agriculture and climate of the area, plus descriptions of each kind of soil.

Laying Out the House

You have a site and you have your house plan. How do you put the pieces together so that your home will take the most advantage of light, air, and view while protecting you from the elements and minimizing heat gain or loss?

The first step is to pinpoint a building site that has firm, well-drained soil. This point can not be overemphasized. Over the life of your home, water can cause critical damage either insidiously or dramatically. Moisture can eat away at building foundations, and flooding can cause severe problems with your septic system and basement.

Visit your property after a rainfall. This will give you a first-hand view of how the site retains water. Check the local building department, health de-

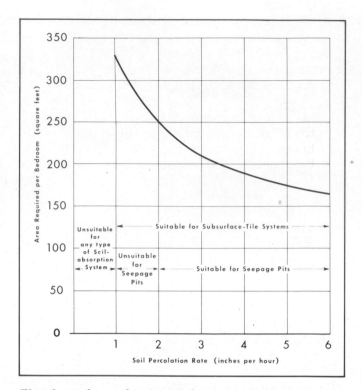

The chart shows the size of absorption fields needed for homes. Chart courtesy of Soil Conservation Service, U.S. Dept. of Agriculture

partment, and/or a soils engineer for more technical expertise.

Assuming that your site presents no drainage problems, here are a few guidelines to assist you in finding the best spot to lay out the home:

- Do not build in low places where air might be trapped.
- Location on top of a ridge will expose your house to higher winds, hotter sun, and noises from a valley below.
- Valleys are generally cooler and quieter as well as closer to water sources. Sloping sites that are well-drained are very good for basements.

The layout of your house will not only affect your views from different rooms and therefore your leisure time ambience but will also influence energy use. Here, then, are guidelines for laying out the house to enhance your comfort and conserve energy at the same time.

- The living room should be oriented to the east or the south to take advantage of the sun for heating.
- The dining area can also be oriented to the south or east. If the room is located on the west side of the house, it would be wise to plant deciduous trees or other kinds of shrubbery outside. This will cut down on glare and overheating during the summer. Building a deck off the dining room facing west will also create a pleasant environment.

- The kitchen should have an eastern or southern exposure to utilize natural sunlight. A kitchen facing north or west will be darker, chillier in the winter, and therefore considerably less cheerful than a kitchen facing south or east.
- Bathrooms can face in any direction; the amount of time spent in the bathroom is relatively short, compared to other parts of the house.
- Bedrooms should be placed facing east or south. During the summer, an east- or south-facing bedroom will not become terribly warm until after you've used the room, and by the time you are ready to use it again, it will have cooled off after several hours in the shade. In the winter, the early morning sun will help heat the room. If your bedroom faces west, you may have to install a fan or air conditioning to cool it before retiring at night.
- If any of your major activity areas face north, consider planting evergreens outside. They will act as a windbreak during the winter.

Piers and Foundations

A house is not built directly on a building site. It must rest on piers or on a foundation for structural strength and protection from the elements. Piers are heavy, vertical support members made of masonry, wood, or metal.

Before you dismiss difficult sites such as waterfront properties, beaches, or steep hilly mountain properties as unbuildable, consider the possibility of building a house on piers.

One form of pier building is pole construction, which offers several advantages over more conventional foundations. It is cheaper, it lends itself to many design possibilities, and it is less disturbing to the environment. Cost savings are available using pole construction, and when properly preserved, pole construction will last for 40 years or more, with great resistance to severe weather.

There are two types of pole construction: pole frame and platform.

In pole frame houses, pressure-treated wood poles are the main vertical members. They extend from below the ground surface to the roof. They support the vertical weight of the building as well as provide lateral resistance to flood, wind, and earthquake. One advantage of this method is that once the poles have been set, the roof can be built before the walls are erected, which makes building partitions, walls, and floors more convenient in poor weather.

On hillsides, a platform can be used, permitting conventional construction methods to be used on the house.

Before planning pole placement and home de-

Steep sites, such as this canyon-side site, can accommodate pole homes. Poles can be installed by pile driving, drilling, machine driving, or by hand. Photo courtesy of Stuart MacArthur Resor, Architect

Advantage of pole construction is that the surrounding environment remains relatively undisturbed, as shown in the construction of this four-bedroom, mountaintop- trilevel. Photo courtesy of Stuart MacArthur Resor, Architect

sign, seek advice from a soils engineer, architect, or builder about your site. Factors to be considered are the proposed size of the house, the steepness of the slope, the type of soil, and drainage patterns on the site. These will affect the types of poles to be used, how deep they should be placed, and what type of backfilling they will need.

Poles can be installed by pile driving, drilling, machine driving or, of course, by hand. The latter is cheapest and disturbs the environment least of all methods. If you are doing it yourself, it is likely to take longer and prove more physically tiring. If adequate pole embedment is not possible, a keywall of concrete should be connected to the frame, preferably at floor level, to reinforce a pole platform.

It is important to try to preserve natural vegetation and drainage patterns around and under the pole house to forestall slides and erosion. A slide can occur, for example, if the backfill around the pole settles, causing the area around the pole to soak up water.

Poles can be backfilled with concrete, soil and cement, sand, pea gravel, or crushed rock. Least expensive is clean sand. Soil and cement make an economical combination. Earth removed from the holes is mixed with cement in a five-to-one ratio, then wetted and tamped into place. Particles larger than 1 inch should be removed from the earth.

When putting poles into a steep site where there is a road on top, place a 2 x 10 into each hole and slide the pole down the hill, butt first, until it hits the 2 x 10; then lift the other end.

To plumb and center the poles, have two people with plumb bobs stand at right angles and 20 feet or so away from the pole. Use 2 x 10s to pry the bottom of the pole into the correct position; then start eyeballing it with plumb bobs. Get the bottom 10 feet vertical; the upper portion can be manipulated into the correct position later. Poles can be framed with pairs of beams, girders, or rafters, one on each side.

One thing to keep in mind in planning a pole house is that your insulation and mechanical systems must be particularly suited for its use. The underside of the house is exposed, which can make mechanical systems more vulnerable to freezing, moisture, and other climatic problems. A few hints are:

- Keep all domestic and drain piping within insulated joist spaces.
- Group piping to drop to grade at one point, back from the perimeter of the structure. An insulated enclosure can protect the drop.
- Put rigid insulation on top of the floor; put another floor covering such as carpeting on top of the insulation.
- Seal undersides of joists, filling the spaces between them with blown material or foil-faced insulation batts.

Precut and prefabricated homes can be placed on platforms, which can also mean some cost savings for you. You do not need a difficult, hilly, rocky, or sandy site for pier construction. Pier homes can be built on flat land, too, which would give you a treetop view of your property as well as more shady outdoor areas to enjoy during hot weather.

If pier construction is not to your liking, however, or if your site is relatively flat and well-drained, then a concrete foundation (either slab-on-grade or with basement or crawl space) would be more suitable.

It is particularly important for the foundation to be built right; if it is not, your house could settle unevenly, which causes cracks.

Before the concrete slab is poured in slab-on-grade construction, all utilities should be placed where they will eventually go. Utilities will be installed before walls and floors are enclosed and finished. Likewise, drains and sewers should be placed before foundation footings are poured.

In crawl space construction, floor joists usually span about half the house's width, with a beam supporting the ends near the center of the house. Smaller joists can be used with shorter spans, with joists supported with two or three beams.

In concrete-slab construction, the thickness of the slab under load-bearing walls can be increased instead of requiring pouring of separate footings.

Consider that chimney or column footings support a heavier load per square foot than sidewall footings. The footings must be sized in proportion to their load; heavy loads require large footings, while lighter loads need smaller footings.

Dealing with Local Building Departments

Throughout the entire building process, you will be dealing with the local building officials. In fact, before you turn over your first shovelful of soil, you will have gone through a process of seeking a building permit. In many communities, even the most rural, your plans will have to comply with local zoning regulations, or you will have to get a zoning variance.

As in most social situations, when working with local building officials, the rules of common sense apply. Be courteous, inquisitive, and respectful. Belligerence, anger, and aggressiveness will not help. Many zoning and building department officials are quite knowledgeable and helpful. Seek their guidance. They can give you invaluable tips not only about what local regulations permit you to do and prevent you from doing, but they can also give you good insight about availability of materials, past building experience in the area, local labor conditions, and more.

Many building departments conduct four inspections during the course of a home's construction. The first encompasses the foundation, prior to laying. The second takes place after the roof is framed and pipes, chimneys and vents have been put in. The third is an inspection of the walls before wallboard has been applied or siding has been put up. Finally, when the house is complete and ready for occupancy, the building department will make a last inspection and hopefully issue a certificate of occupancy.

If you decide to purchase a precut or prefabricated home, do not automatically assume that it will meet local building regulations just because the manufacturer says it will.

Building It Yourself

Building your own small vacation home can be a most rewarding experience. But when things go wrong—and they can—it becomes an overly long, tiresome, and expensive process. In this chapter, we will detail the major steps involved in laying out and building the foundation, the superstructure and the finishing. By the time you finish reading this chapter, you will have a better understanding of what is involved in home construction and whether or not the project is right for you.

Do you really have the time for such a project? Do you have enough money for materials and also for hiring professional help if you need it? Is your interest in the project strong enough to carry you through to completion?

If you have answered "no" to one or more of the above, think about hiring someone to do some or all of the building for you. Perhaps hire a contractor to build the foundation and superstructure. Hire sub-contractors to do the plumbing and electrical hook-ups; then finish the inside yourself.

Further, if you are handy with tools and are confident in your ability but have never built a house before, keep the following in mind: (1) the house will probably take longer to build than projected; (2) it may cost more than estimated.

There are more positive aspects of building a small vacation or leisure house. If you can do the job efficiently and relatively quickly, making few mistakes, project expense will be little more than the cost of materials. If you do not have a great deal of money to put into a second home, this alone could put the project within reach. The gratification of undertaking and completing the project is great. And by building a small house, you will develop a new skill that has its own rewards.

Planning the House

Planning a house takes a lot of time and effort and is detailed in other sections of this book. What is detailed here is a very simple house. It is built with a concrete block crawl space; the walls of the unit are constructed of wood studs and plywood siding. The roof is plywood sheathed and finished with asphalt shingles. Windows, trim, and the finishing details are up to you.

Although there are many other types of materials with which you can work, these were selected because they are relatively inexpensive and are easy to use. Siding, for instance, is available in many other types. But plywood panels can be nailed to stud walls while the walls are still on the flat and then the walls can be tipped into place.

If you are going to build it yourself, keep it small and simple. A small, well-finished unit is much better than a large, overly ambitious shelter that is never quite finished.

Laying Out Foundation

The layout of the foundation is one of the first physical steps in the construction of a house. It is also one of the most important. A mistake made when creating the foundation will add to the cost of labor and materials. Two elements are crucial: (1) you must lay out the foundation on the proper location on the site; and (2) you must lay it out square and level.

Laying out the foundation on the proper location of the site is not as obvious as you may think it is. On small lots, municipalities are very exacting in their requirements. If you have a half-acre vacation lot, zoning may specify that the unit be set back 40 feet from the road and either 20 or 30 feet from the adjoining property. This allows you very little flexibility. If you do not adhere to one or more of these requirements, you may have a very difficult time getting a certificate of occupancy. This becomes far less important if you have purchased half a mountainside or 10 acres of field upon which to build.

After you have taken all zoning requirements into consideration, you need to look at the site from the point of view of drainage, sunlight, grade, trees, and view. Unless you have seen your property in all four seasons of the year, be suspicious of low-lying areas. It could be a floodplain. You may build your house in early summer and autumn only to find it under a foot of water in the spring. A local architect, engineer, or building inspector can be invaluable in helping you select a location if you are unsure. If you have a stream running through your property, you are lucky, but make sure you are safely above the floodplain.

Once a site has been selected, clear the immediate area in and around the proposed foundation. You do not have to defoliate the entire property, just clear the foundation area and the space immediately around it.

If you can afford it, hire a surveyor to lay out the

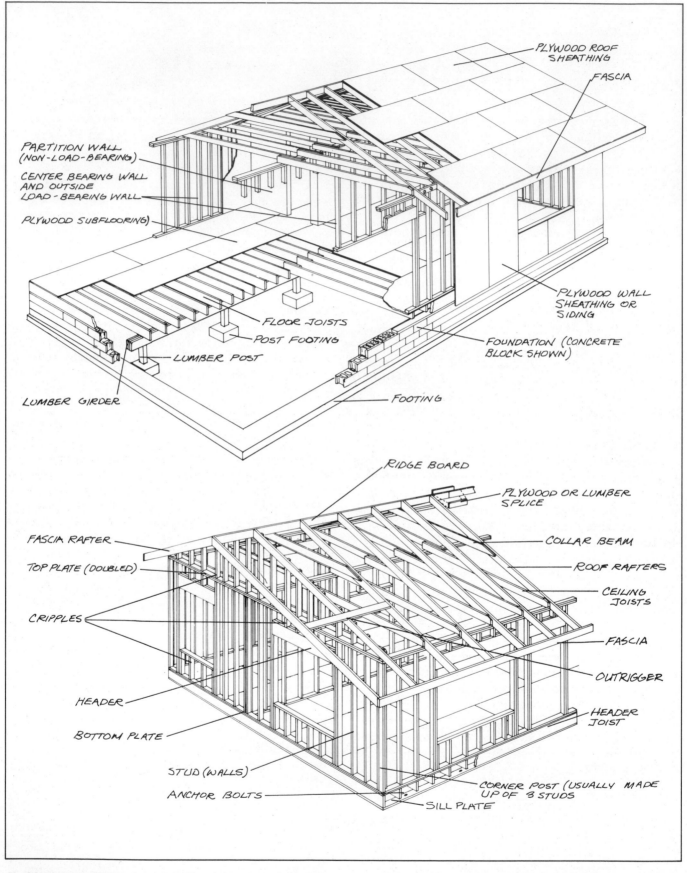

PLYWOOD ROOF SHEATHING

FASCIA

PARTITION WALL (NON-LOAD-BEARING)

CENTER BEARING WALL AND OUTSIDE LOAD-BEARING WALL

PLYWOOD SUBFLOORING)

FLOOR JOISTS

POST FOOTING

LUMBER POST

LUMBER GIRDER

FOOTING

PLYWOOD WALL SHEATHING OR SIDING

FOUNDATION (CONCRETE BLOCK SHOWN)

RIDGE BOARD

PLYWOOD OR LUMBER SPLICE

FASCIA RAFTER

TOP PLATE (DOUBLED)

CRIPPLES

HEADER

BOTTOM PLATE

STUD (WALLS)

ANCHOR BOLTS

SILL PLATE

CORNER POST (USUALLY MADE UP OF 3 STUDS

HEADER JOIST

OUTRIGGER

FASCIA

CEILING JOISTS

ROOF RAFTERS

COLLAR BEAM

Building a small house yourself can be an exciting project. When the project gets too big, it can become overly complicated and costly. Detailed above are the basic elements of a small straightforward dwelling. Drawing courtesy of American Plywood Association

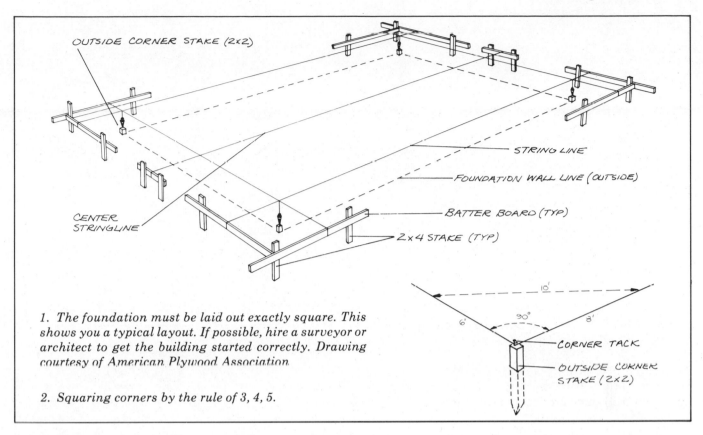

1. *The foundation must be laid out exactly square. This shows you a typical layout. If possible, hire a surveyor or architect to get the building started correctly. Drawing courtesy of American Plywood Association.*

2. *Squaring corners by the rule of 3, 4, 5.*

foundation. He has the proper instruments, and the corners of the house will be laid out exactly square and level.

If you can not afford it, here is how to do it yourself. Locate each outside corner of the house and drive in a 2 x 2 stake (illus. 1). Drive small nails into the top of the stakes to indicate the outside line of the foundation wall (not footings). Check for square.

Measure the diagonals, corner to corner, to see if they are equal. A rectangle will always have equal diagonals if all corners are in square. To check each corner, use the 3, 4, 5 triangle concept (illus. 2). That is, if a right triangle has one side which is 3 feet and another which is 4 feet, the hypotenuse will be 5 feet. You can use multiples of this concept. To check corners, measure down 6 feet on one side, 8 feet on the other and if in square, the hypotenuse will be 10 feet.

Once corners of the foundation are located and exactly square, drive in three more stakes per corner at least 3 or 4 feet outside the perimeter. Do not place them any closer than this, or they will get in the way or become dislodged.

Nail in 1 x 6 batter boards horizontally so that top edges are all level with one another. Tie a string across the tops of opposite batter boards. Using a plumb bob, adjust strings so that the plumb bob just touches the tack which represents the corner of the foundation. Cut "saw kerfs" or notches where the string touches the batter boards. This is done so

that once string has been removed, you can later retie the string without having to remeasure.

Next, find the girder location, which usually runs down the centerline of the house. Check your house plans to determine the exact location. In very small cabins you might not need a center girder. Once you find the girder location, install more batter boards (illus. 3).

Once everything is set up, you must again check for level. A surveyor can handle this quite easily and more quickly than you can. If you want to do it yourself, the best way is to secure a very straight piece of lumber between 10 and 14 feet long. You will use this as a giant ruler or straightedge.

Using the straightedge in conjunction with a good carpenter's level, drive stakes around the perimeter of the house. Resting one end of the straightedge on a batter board, drive in the first stake, at a point on the foundation line which does not exceed the distance of the straightedge.

Now place the straightedge on the first and second stake. Adjust the second stake for level. Continue this process of driving stakes and leveling until you get to the end of the foundation line. With the last stake driven, check for level by placing the straightedge on the last stake and the batter boards nearest to you.

Continue driving stakes and adjusting for level all around the foundation line. Your final check will come when you drive in the final stake, which

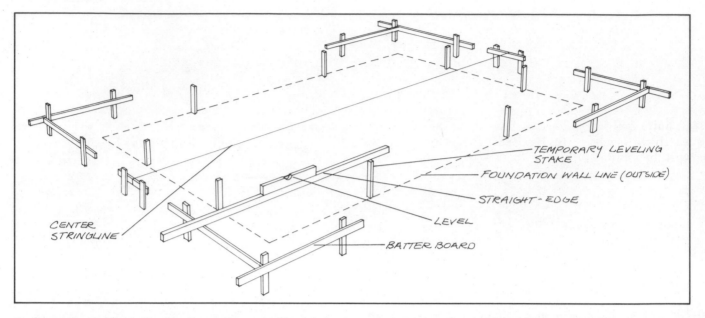

TEMPORARY LEVELING STAKE
FOUNDATION WALL LINE (OUTSIDE)
STRAIGHT-EDGE
LEVEL
BATTER BOARD
CENTER STRINGLINE

3. Once a basic layout is completed, the centerline girder location should be established. When all is ready, ham- *mer in temporary stakes and check carefully for level. Drawing courtesy of American Plywood Association*

should be near the first stake driven. Check these for level. If they are level, then all your batter boards are level. But once again, the most efficient way to accomplish this is using a surveyor's level.

Once excavation gets under way, the corner stakes which denote the foundation corners and the temporary stakes you drove will be removed. Therefore, it is extremely important that the batter boards and string lines be exactly level. Later, when the excavation is checked, the strings will be retied in the saw kerfs.

Footings and Foundation

The next step is to dig the footing trenches. This can be done by hand or by machine. Although a machine is somewhat expensive, you should spend the money, if your labor and time are worth anything at all. Doing it by hand is hard, dirty, and tiring work.

Concrete footings are usually preferred. Properly sized and constructed, they prevent the house from settling, which in turn prevents cracks in the walls. Although your building code will determine what type of footing must be used, typically footings are located about 1 foot below frost line in undisturbed earth. Footings cannot be placed on any type of loose soil because this will not support the tremendous weight of the house.

Footings can be dug in one of two ways. You can dig a relatively small trench and then use the earth walls as a form. Or you can dig a relatively wide trench and then install wooden forms in which to pour the concrete.

If you want to use wooden forms, dig a trench to the specific depth, making it wide enough to climb in to install the forms. The exact width will not affect anything, but the depth will. If you have a frost line of 2 feet in your area, dig the footing trench 3 feet deep. If you dig deeper than you intended, you will have to fill the gap with concrete, not loose earth.

Once the trench has been dug, rehang the strings on the batter boards and, using the plumb bobs, locate the corners of the foundation-to-be. When you start to lay the concrete blocks, they will have to sit in the center of the footing. The footing is wider than the foundation block, therefore when you determine the actual corner, you will have to measure out a few more inches to locate the footing line.

Here is an example: you are using 10-inch concrete block for your foundation wall, and your footings will be 18 inches wide. Once you determine the corner of that foundation wall, the footing will extend another 4 inches on either side of the foundation wall. Thus, if you had your footing in place and centered the block, you would see—looking straight down on it—4 inches of footing on the outside, the 10-inch block, and 4 inches of footing on the inside (illus. 4).

The footing must be outlined on both sides of foundation wall. You can do this with a string, or you can do it with a powdered chalk. The footing forms can be made of either lumber or plywood. Select a sturdy but inexpensive material because the material used is such a mess by the time you finish that it must be thrown away.

If you use plywood, you can cut a 4 x 8 sheet into 1-foot widths and, using stakes, install it in the trench. If possible, try to dig the original trenches narrow enough so surrounding earth can be used as a form. Remember, the trench must be exact.

With the trenches and forms ready, you can now dig the footings for the supports which will hold the centerline girder in place. These holes are called post footings and are between 20 and 24 inches square (check codes and plans). Usually, it is easier to build box forms for this pouring rather than use earth walls as a form. Boxes for these post footings can be made similarly to the above mentioned forms.

When forms are all ready, check your plans to see if you have to install steel reinforcing bars. Many codes demand them. They tie the entire footing together.

When ordering concrete, estimate carefully how much you will need. It is better to err on the side of too much rather than too little. Concrete is ordered in cubic yards. One cubic yard contains 27 cubic feet of concrete. If you were to pour a footing which is 1 foot deep by 1 foot wide, you would get 27 running feet of footing from a yard. You can figure out how much concrete you need by calculating the dimensions of your footing. If, however, you are using earth sidewalls as a form, order a few more yards because the excavation will never be exact.

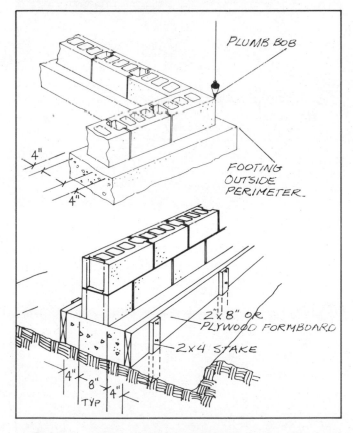

4. *Before mortar is placed on the footing, place blocks in position to assure proper fit. The block should be positioned in the middle of the footing for best results. Drawing courtesy of American Plywood Association*

For concrete, have extra help at the building site. The material is too heavy and the work too exhausting for one person to do alone.

Never pour concrete if you expect temperatures to drop below 40 degrees Fahrenheit during the next week. If temperatures do drop lower, you will have to take precautions to keep the material above that temperature.

Codes and your building plans have the final say, but usually you need a concrete mix of at least 2,000 psi, 28-day strength.

To place concrete in footing forms, shovel it in by thin layers. Rake and tamp to remove all air pockets. Continue this process until forms are filled. Tops of footings should be smooth and level all the way around. Use a trowel to adjust surface.

Allow several days for the footings to cure. Concrete does not dry, it sets. When it has set, remove the wood forms, if they were used.

Laying the First Course of Blocks

Whether your house is a slab-on-grade with crawl space or a full foundation, the procedure for laying the first course of blocks is the same.

First, find the outside corner of the foundation wall as described earlier. Retie strings to locate the corners. Outline where the blocks will go; use either a piece of chalk or snap a chalk line on the footing.

Lay the bottom course of blocks completely around the perimeter of the footing without mortar. This way, you will be able to see where you have to cut blocks to create a good fit. Generally, blocks should be placed an average of ½ inch apart.

With blocks resting in place without mortar, mark the joint locations on the footings with chalk. Remove the blocks. Before you actually lay the blocks with mortar, double-check to make sure where all plumbing and utility openings will be.

For laying block, a mortar mix is prepared using two parts masonry cement (or one part each of Portland cement and hydrated lime) with four to six parts of damp mortar sand. Water should be added slowly. The consistency should be such that the material clings to the trowel but does not squeeze out when a block is placed on it.

Lay blocks as shown (illus. 4). To begin, trowel the mortar on the footing and place block on top. Work it into the masonry and check for level. Place the next block guided by the chalk mark on the footing. Adjust and level. Now trowel mortar into the joint between the two blocks and scrape off any excess. For succeeding courses, you will only place mortar on the face of the block.

When adding each new course of blocks, stagger

the joints. The wall is not built adding on one course upon another; rather, you must build the corners up first to full height to establish the required thickness of joints. Use corner blocks with one flat end at corners. Use a mason's level to keep all blocks plumb and level. Stretch a line between corners to guide the laying of additional blocks.

Before finishing the last two courses of blocks, locate and position anchor bolts as shown on your plans (illus. 5). You will have to provide at least two bolts per individual sill plate. Fill all cells in the top course with mortar. When the wall is completed, wait at least a week before you backfill against it. Install 2 x 6 sill plates around perimeter of the foundation.

If this seems too much trouble, you might want to investigate a poured concrete foundation wall. It can be used for crawl space, slab-on-grade, or for full foundation walls, and is a lot of trouble to do yourself. You might want to locate a contractor in your area who specializes in this work. He will supply concrete forms, labor, and materials. You will save a lot of time and be able to get on with the building of the superstructure more quickly. Depending on your area, a poured concrete foundation is usually price competitive with a block foundation. Get bids from at least two contractors and compare costs yourself.

5. Here are some tips on the proper construction of a concrete block wall. Drawing courtesy of American Plywood Association

Posts and Girders

Vertical posts run from the boxed footings and support the horizontal girder. The girder, along with the foundation walls, supports the floor joists which in turn support the interior flooring and load-bearing walls.

Posts are constructed of lumber which are solid pieces of wood either 4 x 4 or 6 x 6. Girders are generally constructed of either two or three pieces of dimensional lumber such as 2 x 8, 2 x 10, or 2 x 12.

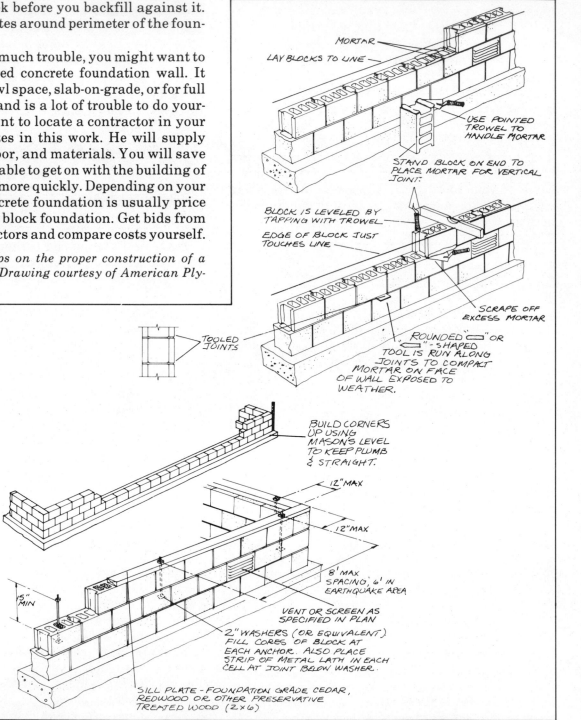

MORTAR

LAY BLOCKS TO LINE

USE POINTED TROWEL TO HANDLE MORTAR

STAND BLOCK ON END TO PLACE MORTAR FOR VERTICAL JOINT.

BLOCK IS LEVELED BY TAPPING WITH TROWEL

EDGE OF BLOCK JUST TOUCHES LINE

SCRAPE OFF EXCESS MORTAR

TOOLED JOINTS

ROUNDED "⌐" OR "⌐"-SHAPED TOOL IS RUN ALONG JOINTS TO COMPACT MORTAR ON FACE OF WALL EXPOSED TO WEATHER.

BUILD CORNERS UP USING MASON'S LEVEL TO KEEP PLUMB & STRAIGHT.

12" MAX

12" MAX

8' MAX SPACING, 6' IN EARTHQUAKE AREA

VENT OR SCREEN AS SPECIFIED IN PLAN

2" WASHERS (OR EQUIVALENT) FILL CORES OF BLOCK AT EACH ANCHOR. ALSO PLACE STRIP OF METAL LATH IN EACH CELL AT JOINT BELOW WASHER.

5" MIN

SILL PLATE - FOUNDATION GRADE CEDAR, REDWOOD OR OTHER PRESERVATIVE TREATED WOOD (2 x 6)

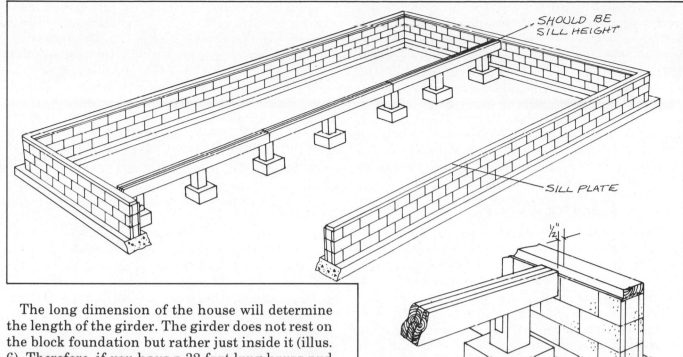

The long dimension of the house will determine the length of the girder. The girder does not rest on the block foundation but rather just inside it (illus. 6). Therefore, if you have a 32-foot long house and you are using 8-inch block, the length of the girder will be 32 feet minus 16 inches.

The girder, as a main support of the house, must rest squarely on the posts. In the 32-foot house mentioned above, the center girder would be made in two 16-foot sections using three 2 x 12s nailed together with 20d common nails 32 inches on center in each of two rows—one along the top and one along the bottom of the girder. Stagger the top and bottom rows (illus. 7).

Once the girder is made in sections, you must determine the exact length of the posts. Stretch a line from the sill plate across the designated position for the girder to the other sill plate. Use a strong line which can be stretched tightly. When the string is taut, measure the distance to each footing; then subtract the dimension of the girder. That is the length of the post. You must make separate measurements for each post because there may be a slight variation. If post footings have a protruding piece of reinforced bar, drill the post bottom so that the post can slip over the bar.

Sometimes building codes insist on the installation of a vapor barrier in the crawl space. If this is the case with your house, install it now and cover the post footings with the material. If not, place a piece of 15-pound asphalt-impregnated building felt between concrete post footing and the post end. Posts will rest on this material (illus. 8).

Once posts are positioned, lift the girder and place it on top of the posts. The girder must be cut so that the butt joint falls over the center line of the supporting post (illus. 8). Trim the two end girder sections to allow for a 1-inch clearance. Unless you are going to

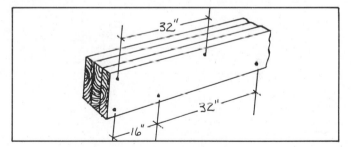

6. *The girder does not rest on the block wall but rather just inside it. Girder butt joints should only occur over vertical support. Drawing courtesy of American Plywood Association*

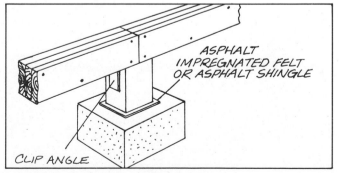

7. *Construction detail for making a support girder. Drawing courtesy of American Plywood Association*

8. *If a vapor barrier is not specified by codes, place a piece of 15-pound asphalt-impregnated building felt between concrete post and footings. Drawing courtesy of American Plywood Association*

immediately install floor joists, brace posts and girder sections with 2 x 4s to hold them in exact position.

Now check for level. The entire girder must be level with sill plates. If there is a dip or an unlevel spot at one post position, shim between girder and post with a piece of cedar shingle. Once level, connect girder and posts. Toenail at least six 10d common nails into the posts. Firmly attach the girder to each post on the underside with galvanized steel framing anchors or clip angles which are available from a well-stocked hardware store.

Laying Floor Joists

With the foundation complete and sill plates and girder in position, you are ready to begin on the superstructure. First, floor joists must be laid. The joists on this level are really the main support of the house. They rest on sill plates and the center girder.

Typically, floor joists are either 2 x 8, 2 x 10 or, in special situations, 2 x 12. They are positioned either 16 or 24 inches on center.

Before this portion of the work begins, check plans and building codes to determine if a special type of wood grade is called for.

Often you can save money here by placing the joists 24 inches on center and then using a thicker plywood subfloor. Check with the building inspector.

Once your lumber is on site, notice that most joists have a bow and a crown on them; this means that on one edge the lumber is usually slightly warped one way or the other. You must always nail the side that warps up in an "up" position. Once the floor is in place it will straighten out. If you place the joist with the crown down, you will have a dip in the floor after it is installed.

Joist layout depends on plans, but here are several steps to aid you in understanding joist layout.

1. Begin layout by starting at the edge of the sill plate (illus. 9). Use the outside of the sill plate as a starting point. Because joists must be located 16 inches on center, unless you are using a 24 inch on center layout, you must calculate the thickness of the joist, then mark on the sill plate where the joist would touch. Find the center between the two marks and measure in 16 inches. This would be the point where the center of the second joist would touch. Continue this along the wall until you reach the end of the house.

2. Once the first side is marked off, you must then mark off the parallel wall in the same manner. All joists must be parallel to one another. Now mark correct joist locations on the center girder.

3. Study the floor plans very carefully now. When

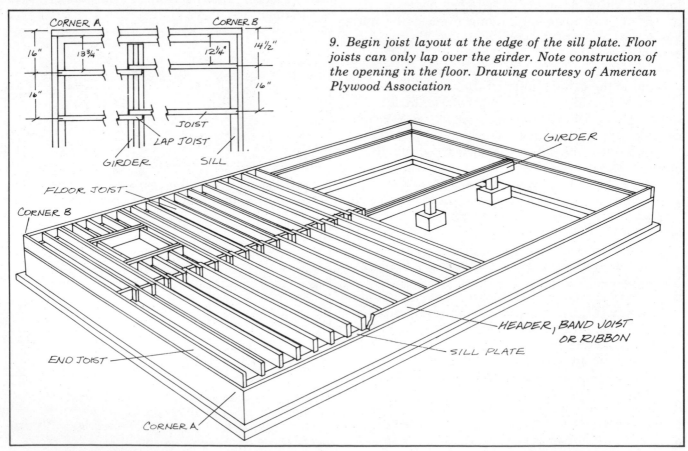

9. Begin joist layout at the edge of the sill plate. Floor joists can only lap over the girder. Note construction of the opening in the floor. Drawing courtesy of American Plywood Association

This beautiful year-round vacation home features huge windows in the family room and multiple deck areas. Photo courtesy of Pan Abode Cedar Homes

Redwood is the primary building material in this coastal vacation home. The deck area has almost the same square footage as the home. Photo courtesy of the California Redwood Association

Piers which follow the contour of the land can be economical. Minimal site disturbance was necessary for the construction of this lakefront leisure home (opposite). Photo courtesy of American Plywood Association

This oak log two-story home is set off in winter by evergreens. The home has porches, front and rear, running the full length of the house.

A geodesic dome is an easily constructed and exciting leisure home. There are a variety of kits on the market for the do-it-yourselfer. Photo courtesy of Domes and Homes, Inc.

When the budget permits, solar heating and hot-water systems are a nice addition to a leisure home. Photo courtesy of Acorn Structures

Natural wood siding and judicious placement of windows make this retreat a lovely sight to behold from both inside and out. Photo courtesy of Acorn Structures

Breathtaking views of the ocean from house and deck make this dwelling a year-round retreat. Photo courtesy of Karl Riek

A simple packaged home, built on piles, can offer extra storage space beneath the home. Photo courtesy of Jim Walter Homes

Good design captures light and space for this vacation home. Photo courtesy of Michael Zide

The hexagon is one of the most space-efficient shapes, is easily fabricated, and is versatile enough to offer years of pleasurable living. Photo courtesy of Western Wood Products Association

Small, compact, and easily constructed are the virtues of this lakeside home. L-shaped deck is situated to take in both land and lake views. Photo courtesy of Jim Walter Homes

This contemporary design (lower right) with a "crow's nest" is indicative of what is available from manufactured housing producers. Photo courtesy of Acorn Structures

Built from scratch, this home of 6-inch, squared-on-three-sides logs features a steep roof line and double sliding glass doors that open onto a deck.

This cozy small cabin has a log base and planked gables with a large loft area in the attic. Photo courtesy of Beaver Log Homes

The design should fit purpose and need. A basic small cabin (opposite) is ideal for a few weekends each year. Photo courtesy of Pan Abode Cedar Homes

A vacation home can be as elaborate as you desire. This comfortable two-story features a hot tub sunken into the deck area. Photo courtesy of Pan Abode Cedar Homes

This beautiful two-story log home was constructed with logs and stone found on or near the building site.

Joist construction for most log homes is similar to the above layout.

This close-up shows a good method for joining round logs at the corners. Photo courtesy of Rocky Mountain Log Homes

A vacation home should conform to the surrounding terrain. Windows in the loft of this home overlook the surrounding land. Photo courtesy of Alta

A ceiling fan adds charm and helps keep this simple log cabin warm by preventing heated air from rising. Photo courtesy of Beaver Log Homes

This attractive bedroom has access to an outside deck through sliding glass doors. Photo courtesy of Pan Abode Cedar Homes

Cabinets in this kitchen were constructed by the owner to achieve greater savings.
Photo courtesy of Pan Abode Cedar Homes

This country kitchen has all the conveniences of home. The skylight permits natural light to enter the room. Photo courtesy of Pan Abode Cedar Homes

The kitchen below features standard cabinet units and an informal dining table. Photo courtesy of Pan Abode Cedar Homes

A simple, efficient kitchen can be created by extending a work area from any convenient wall. Photo courtesy of Pan Abode Cedar Homes

A den was created in the attic of this well-designed home. Photo courtesy of Pan Abode Cedar Homes

Simple staircases are easiest to build and are more efficient than spiral staircases.

Even spiral staircases are available for log homes. Photo courtesy of Pan Abode Cedar Homes

Most multiple-story log cabins have open loft areas. Photo courtesy of Pan Abode Cedar Homes

Whether renovated or newly constructed, a spacious barn makes an ideal vacation retreat for that individual or family who needs lots of inner space. This barn was renovated to an architect's design and is in the heart of Napa County, California. Photo courtesy of Karl Riek

This bathroom is a far cry from the outhouse our ancestors used. Features include an adjacent sauna and whirlpool. Photo courtesy of Pan Abode Cedar Homes

Certain design elements, such as this contemporary outhouse-style bathroom, can be traced to the first log cabins.

A two-story fireplace can transform a "log cabin" into a rustic home. Photo courtesy of New England Log Homes

This photo shows rafter and roof joist details.

Open log rafters and tie beams particularly suit log construction. Note the pole posts and handrail on the balcony. Photo courtesy of New England Log Homes, Inc.

you come across an interior wall which is running parallel to the floor joists, the wall not only needs to be located on top of a joist but it also needs a double joist positioned to help support this extra weight.

4. At this time also check plans for openings in the floor. This would include openings for a crawl space or an opening to the basement, if you have a full foundation. You will lay all floor joists the same, but you must be aware of where openings will be.

5. Now install end joists and headers. Save your straightest pieces of lumber for this. End joists and headers form a box around the remainder of the floor joists. The headers are the beams which fit perpendicularly at the end of the floor joists to form that box. Now nail it off. Use large 16d nails to attach the end joists to the headers. At intervals of 16 inches, toenail the headers and end joists into the sill plates.

6. Now you are ready to install the other floor joists. Because you have marked off the sill plates, you know the location of each floor joist. To ensure that each joist is square on the vertical, use a steel square and scribe a line on the inside of the header.

7. Cut to size and install joists. If lumber does not span the entire dimension of the house, then it must overlap over the center girder (illus. 9).

8. Joists can be secured using two or three 16d nails. Nail through the header and into the floor joist. Toenail the joist at the girder. If joists overlap at the girder, nail them together and also toenail into the girder. The overlap should be a minimum of 4 inches and a maximum of 24 inches.

9. Find the floor openings, once all joists are secured. Cut and frame all openings as shown (illus. 10), or as per house plans. Take care to find the position and cut joists without knocking them out of alignment. Use a double secondary header as shown and end nail into joists.

10. Floor joists need blocking. Use pieces of lumber the same dimension as floor joists; fit them between joists at girder position (illus. 11).

Installing Subfloor

In a typical installation, plywood subfloor is attached to the floor joists. If hardwood flooring is planned, no additional plywood is needed on the floor. If tile or carpet will be applied directly to the floor, another layer of plywood is needed. This second layer is not added until much later in the construction process.

If your foundation is square and has properly installed floor joists parallel to one another, apply-

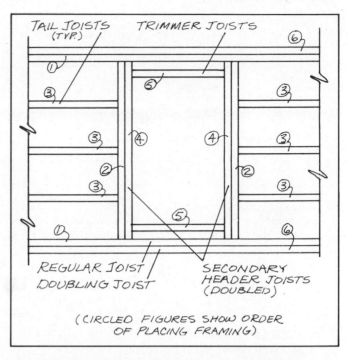

10. Typical framing detail of an opening in the floor. All joists are positioned and secured. They are then cut, and the opening is framed out. Drawing courtesy of American Plywood Association

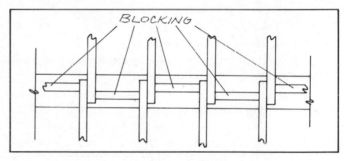

11. Once floor joists are in place, they must be blocked. Blocking is merely pieces of lumber secured between joists. Here blocking is placed on the center-line girder where joists overlap. Drawing courtesy of American Plywood Association

ing the subfloor is easy. Use 4 x 8 sheets of plywood for this job. Before you cut anything, make a small sketch of the plywood layout so you will know what has to be cut.

Start your layout at the same corner where you began laying the floor joists. The subfloor will run perpendicular to the floor joist layout. Measure in 4 feet on the two end joists; then snap a chalk line which cuts across all joists.

Without cutting anything, lay a line of 4 x 8 plywood panels between the chalk line and the outside header. Leave a space of about 1/16 inch between each panel for expansion. You can calculate this as you go along by fitting a dime between panels. Repeat this until you come to the end of the house. If the last panel overlaps the end joist, you can simply

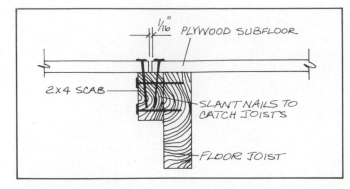

12. If something is out of square and the plywood panel does not rest securely on the joist, then a piece of 2 x 4 can be nailed to the joist. If the plywood is substantially short, the panel should be cut to fit the closest joist. Drawing courtesy of American Plywood Association

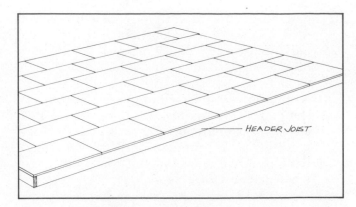

13. Subfloor panels must be staggered. Also, leave about ⅛-inch space between courses of the subfloor so that the wood has room to expand. Drawing courtesy of American Plywood Association

cut it off. If the last panel falls short an inch or two, you will need to install a piece of blocking underneath. Simply hammer a portion of 2 x 4 on the end joist so that there is something to which you can attach the subfloor. The end of the plywood always must be supported underneath (illus. 12).

In the initial layout of the subfloor, each panel should rest on the center of the floor joist. If it does not, you have not installed the floor joists properly.

Once this first section of subfloor is in place, you can begin the second course. The subfloor panels must be staggered (illus. 13). Leave about ⅛-inch space between courses of the subfloor.

Once all panels are in place, you can nail off the subfloor. Probably the easiest way to do this is to go back over your work and accurately snap a chalk line over the location of each floor joist. This will save you time finding the floor joist as you nail and, perhaps, keep you from missing the joist altogether. Using either 6d or 8d nails, nail off the panels. At the ends of panels, drive about 10 nails into it. At intermediate stages, drive about five or six nails.

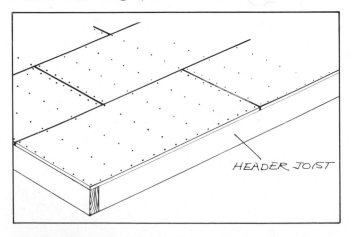

14. Typical nailing pattern for applying plywood subfloor to floor joists. Drawing courtesy of American Plywood Association

Where panels touch headers, space and drive about 17 nails (illus. 14). Take care to angle each nail into the joists and headers. Once this platform is in place, you can begin work on the sidewalls of the house.

Framing the Walls

Wall framing includes installation of vertical wall studs, horizontal members called bottom and top plates, window and door headers, and interior walls.

The most efficient way to build a wall today is to construct it on a flat surface and tip it into place. To do this, you really need at least one helper, perhaps more. If there are only two people building walls, do not build one longer than 24 feet or it will be too unwieldy to tip into place. On a small vacation house, it is doubtful that you would need a wall any longer than this.

Find the location of every exterior and interior wall on your plans and nail bottom plates in those locations. A bottom plate is the bottom horizontal support, made of 2 x 4s, to which the wall studs are nailed. Temporarily nail all bottom plates in proper location. Do not use many nails because you will have to remove the plates later. By determining each wall location, you can figure out where special framing is needed.

With bottom plates in place, begin with exterior walls and find the location of each wall stud. These are placed 16 inches on center. You can determine stud location as you located floor joists. Measure in 16 inches from the end of the wall. Figure the width of the stud and mark it. Find the center between the two marks. From that center measure another 16 inches. This next mark will be the center of your second stud. Continue to the end to find all stud locations (illus. 15).

At the end of walls and at intersections, you will

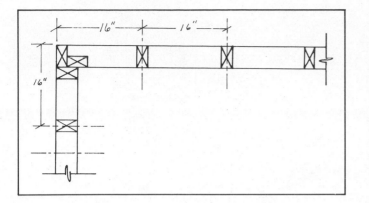

15. *Pattern for wall stud layout. Drawing courtesy of American Plywood Association*

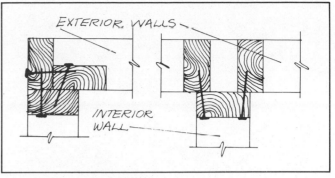

16. *At walls and intersections, a second stud must be added. Here are typical layouts. Drawing courtesy of American Plywood Association*

need to add a second stud. Check your plans for the location of this second stud. When calculating stud locations, measure from the outside stud, not from the extra stud.

Upon completing this, cut another series of plates. These will be the bottom section of the top plates. Top plates are the horizontal members which run along the top of the stud wall. Standard walls have one horizontal bottom plate, vertical wall studs, and two top plates. Once cut, you can place the top plates on top of the bottom plates and keep them together temporarily.

Generally you can buy precut studs. They save time and avoid wasted materials. If you must cut studs from larger 2 x 4s, here is how to do it. Note from your plans the floor to ceiling height of walls. To this add about 1 inch for underlayment. Then subtract about 4½ inches, which is the thickness of three plates.

Carefully cut a stud to length. You can now use this as a pattern for cutting the remaining studs. The next step is to mark on the subfloor the location of every stud wall with chalk. Using the bottom plate as a guide, mark stud locations on top plates. Next remove all plates from a portion of the floor so that you have a place to work. When removing plates, do not mix up the walls. Organize them so that when you finish one wall you can begin another.

Begin building the long exterior walls first. Lay the bottom plate on edge. Lay the top plate parallel to the bottom plate about a stud's distance away. Fill in between the plates with as many studs as necessary. All lumber should be on edge. Holding a stud in its exact location, nail through the bottom plate, then nail through top plate into stud ends. Use two 16d nails for this. The second top plate will be added later.

Add extra studs where needed. If the wall section has a window in it, you should now install header and supports as specified on the plans. Make sure

the wall section is square by measuring diagonals. Remember, if the wall is in square, the diagonals will be equal.

Plywood Walls and Siding

You have several options. If you are going to use a siding other than plywood panels, you can now add sheathing, then tip walls into place. (Siding will be added later.) It is recommended, however, that you use plywood panel siding. You can save the cost of sheathing walls because a single plywood siding panel will do the trick; you can also save the labor of siding later.

Using plywood panels in this way is called single-wall construction. Usually you need a substantial panel for this operation, such as ⅝ inch thick. Both your plans and/or the building code will be the determining factors.

The American Plywood Association offers the following suggestions for installing plywood siding panels in the single-wall framing method.

First, siding panels must be cut to proper length. Be careful in determining length so that you do not waste material. You must allow a minimum of 1-inch lap over the top of the foundation wall and 1½ inches for covering the second top plate.

Place the first panel at one end of the stud wall while it is still on the flat. Make sure the edge of the panel is flush with the outside edge of the corner stud.

Apply the panel to the wall framing. Use hot-dip galvanized, aluminum, or other nonstaining nails to prevent stain from forming on the siding. Use 6d box, siding, or casing nails for plywood siding that is ½ inch or less in thickness; use 8d nails for thicker panels. Drive nails every 12 inches at intermediate supports. All edges of panel siding must be backed by solid lumber framing and blocking.

Between panels, leave a ¹⁄₁₆-inch gap for expansion. If you force panels in close together, they can buckle. Once the wall is sided, tip it into place. You

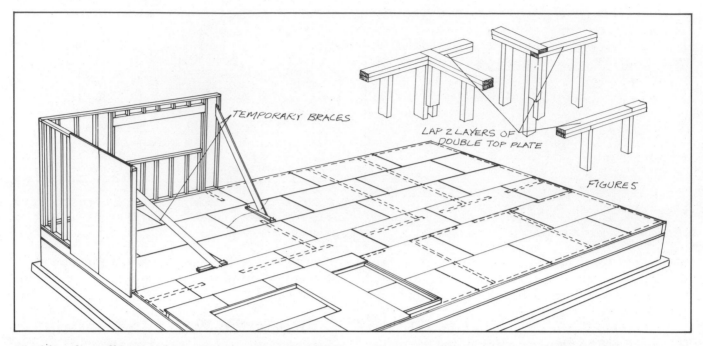

TEMPORARY BRACES

LAP 2 LAYERS OF DOUBLE TOP PLATE

FIGURE 5

17. *After the wall is tipped into place, temporarily hold it in place with 2 x 4 lumber. Drawing courtesy of American Plywood Association*

need help to do this right; you could lose control of a heavy wall when you are tilting it into place. With the wall tilted and steadied, nail through the bottom plate into the subfloor and header or end joist. Temporarily hold this wall in place by using 2 x 4 bracing which can be nailed to one of the wall studs and toenailed into the subfloor with the aid of a small block (illus. 17).

Other long sidewalls can be constructed in a similar manner. When it comes to adding the walls which make right angles with the longer walls mentioned above, a different procedure is necessary. The plywood panels of those walls do not butt the end as above. The panel must extend past the end of the wall in order to cover the end of the larger wall (illus. 18).

The best way to handle this is to build the smaller wall and tip it into place and fasten it. Then apply the panel. Usually you can purchase and install corner trim pieces in case the fit is not exact.

When walls are in place, joints should be caulked. You do not have to caulk shiplapped joints, but you do have to caulk butt joints at inside and outside wall corners. Use a good grade caulking material for this so that you do not have to recaulk it in another year.

Once all exterior walls are in place, you can build the interior walls the same way. They are not covered with anything at this point. Bare stud walls should be built and installed.

After you are sure every wall is plumb and square, you can add the second set of top plates.

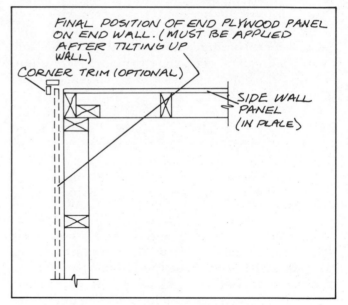

FINAL POSITION OF END PLYWOOD PANEL ON END WALL. (MUST BE APPLIED AFTER TILTING UP WALL)

CORNER TRIM (OPTIONAL)

SIDE WALL PANEL (IN PLACE)

18. *This construction detail shows the attachment of the sidewall to the longer wall. Drawing courtesy of American Plywood Association*

Framing the Ceiling

In a simple house, ceiling joists are installed similarly to floor joists. They span from one exterior wall, over load-bearing interior walls, to the other side of the house. There are many different design motifs, including sloped roofs, cathedral ceilings, and others. Here, a flat ceiling is used.

Ceiling joists have several functions: tie in the walls, offer a structure upon which to apply ceiling material, and support a second-story floor in a two-story house.

Although we are only describing a single-story

vacation house here, if you add a plywood subfloor on top of the ceiling joists you will have extra storage space. At some point you might also want to add dormers and create an extra sleeping space.

The only difference in construction between laying floor joists and laying ceiling joists is that you do not use headers with the ceiling joists. Joists are positioned as per your plans and building codes. Typically, floor joists are positioned 16 inches on center. Many codes allow ceiling joists to be positioned 24 inches on center.

Except for short spans, ceiling joists need support. Support is usually provided by load-bearing interior walls (walls which run perpendicular to floor joists). Where there are broad expanses of space, a beam must be provided for ceiling joist support. Plans will detail this. If a beam is indicated and it is located below ceiling joists, the joists rest on it. If it is on the same level or above, the joists are hung from the beam with the help of metal hangers. This would be specified on plans.

For a small house, you can usually secure lumber which will span the entire distance. If not, then joists must lap over an interior load-bearing wall and nowhere else. If they do overlap, then consider roof rafter position immediately. Roof rafters must always frame opposite each other, but this would be impossible to do with a simple joist overlap.

Lay out the ceiling joists and place a block, the same dimensions as the roof rafters, between the joist overlap (illus. 19). The roof rafters can then be installed on either side of the ceiling joists and thus frame opposite each other over the filler block.

Because the end roof rafter must sit flush with the outside wall, it will not lap with the end ceiling joist. Therefore, it is usually easier to fit the end joists in after the gable end has been framed out. Start your ceiling joist layout approximately 24 inches in from the end, if you are locating ceiling joists 24 inches on center (illus. 20). Continue placing them at specified distances until you get to the other end. You can calculate ceiling joist positions in a manner similar to the way you calculated floor joist positions.

When measuring and cutting the ceiling joists, measure from the outside of the exterior wall to the center interior load-bearing wall or beam. Leave an extra 4 to 6 inches for an overlap with the filler block and the other joist. Place joists with crown up. Toenail joists into the exterior top plate and the interior wall top plate. Use 10d nails for this. With 16d nails, nail through the joist and into the filler block. Do the same with the other joist meeting the filler block so that you have a strong connection. If the joists touch any other interior wall, toenail into that as well.

Access to the attic must be cut and framed out similarly to openings through the main floor into

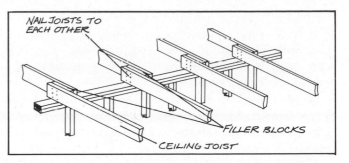

19. *Lap ceiling joists over support. Drawing courtesy of American Plywood Association*

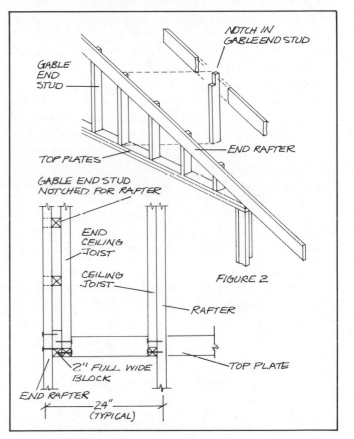

20. *Begin ceiling joist layout approximately 24 inches in from the roof end. The end joist is added later. Drawing courtesy of American Plywood Association*

the crawl space or basement. Eventually the ends of the ceiling joists will have to be trimmed to fit the slope of the roof rafters, but you can do that after the rafters are installed.

Sheets of plywood should be temporarily fastened to the top side of the ceiling joists so that you have solid footing for installation of the roof rafters.

Installing Roof Rafters

Up to now the building of this small house has been straightforward and relatively simple. Cutting and installing roof rafters is probably the most complicated part of building a house. You should not undertake this part of construction without the help of at least one other person.

Framing out the roof not only involves construction and placement of roof rafters, but also construction and placement of cripple studs, ridgeboard, and collar beams.

Roof rafter locations should be determined at the same time that ceiling joists locations are set. If you have done this, then the exact position of roof rafters (to fit over the filler block) will be apparent.

Cutting and placing roof rafters is made somewhat complicated because you have to take the slope of the roof into account. Your home plans will detail this slope for you. Usually the rise is between 4 to 6 inches per foot. On a house which is 24 feet wide, the peak would occur at 12 feet. Multiplying 4 inches by 12 feet, you can calculate that the roof rises 48 inches.

The real difficulty comes because the roof rafters must be cut at an angle that properly meets the ridgeboard (main beam running the length of the house). You must also notch-cut the rafters so that they fit into and are seated with the exterior wall top plate. Finally, you must have the proper overhang, as specified on your plans.

The best way to begin is to draw a full-scale sideview of the rafter construction right onto your plywood subfloor. Set it up so that you take the ridge-

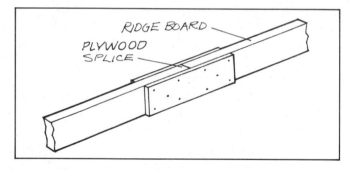

21. *Typical attachment for the ridgeboard. To join two portions of ridgeboard, make a plywood gusset and nail as shown. Drawing courtesy of American Plywood Association*

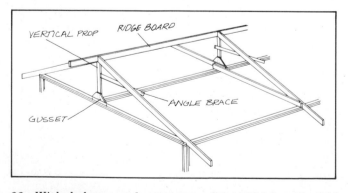

22. *With helpers and supports, place ridgeboard. Add several pairs of rafters to help hold it in place. Drawing courtesy of American Plywood Association*

beam, top plates, and slope into account. Once this is done, take two of the pieces of lumber you will use for roof rafters and lay them on top of the drawing. Now draw in the cuts you will have to make. Set them up on the subfloor in the same manner you would on the roof to see if they fit together properly. Once you have a good fit (don't forget the width of the ridgebeam in this test), you can use these two rafters as patterns upon which to cut the remaining rafters.

If your house is perfectly square, you can cut all the remaining rafters. If everything is not exact, you may want to cut just a few at first to help support the ridgebeam. You can now fit the pattern rafters in place; if they fit exactly, you can then cut more rafters. If they do not, you will have to adjust your cut as you go along the length of the house.

The ridgeboard must be fabricated. In a very small house, the ridgeboard can be one piece of lumber. In most cases, you will have to splice together several pieces. Check your plans to determine what size beam you need for the ridgeboard (illus. 21). Select your straightest pieces of lumber for the ridgeboard. Construct it in several sections, whether on the ground or on the subfloor platform. Take the sections, some 2 x 4 supports, and several sets of roof rafters up to the top of the house. With one or more helpers, nail the temporary support in place at the end of the house on the top plate. Place another support in about the distance of the first length of ridgeboard and nail it to a ceiling joist.

With the temporary supports nailed at the bottom, check the ridgeboard section for level. Firmly hold the ridgeboard while assistants toenail it to the temporary supports. Also at this time install enough diagonal bracing to ensure that the ridgeboard will not move. Now double-check for level (illus. 22).

Install the first set of rafters at the end of the house by toenailing them into the ridgeboard and into the top plate. Install the next set of rafters in a similar fashion near the end of the ridgeboard section. Install the next ridgeboard section and rafters. Continue to the end of the house.

The ridgeboard sections must be joined securely. Make plywood gussets (one for each side of the joint). They should be the width of the board and have enough overlap to enable you to nail it securely (illus. 21).

Again, check the ridgeboard for level and be sure the board is centered over the house. Now install the remaining rafters in pairs, continuing to check for level and straightness. This cannot be stressed enough. A crooked ridgeboard can ruin the whole look of the house. When installing rafters, never force one against the ridgeboard; this could throw it off center. Rafters should be nailed to the top plate with 10d nails. Then using 16d nails, nail rafters to

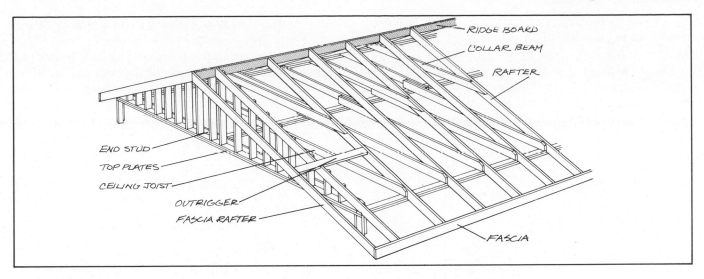

23. *Cut and nail collar beams in place. Drawing courtesy of American Plywood Association*

the ridgeboard. Also nail rafters into floor joists.

Cut and nail 1 x 6 collar beams in place for every other roof rafter in the upper third of the rafter (illus. 23).

If vents are to be installed at the ends of the roof, find the center line from the ridgeboard to top plate. Measure the vent to be installed and leave half this distance on either side of your mark. Install your first set of studs taking into account the width of the vent. Continue to install studs 16 inches on center. The top end of the stud should be cut to fit under the rafter. The stud bottom should fit flush with the top plate. Cut and frame out the vent opening (illus. 24).

Install outside ceiling joists on the inside of the end stud wall. Install fascia board, which will correct the length of the ridgeboard. Check your plans on this, then install fascia rafters to cover the end of the ridgeboard.

Installing Roof Sheathing

Installing roof sheathing is no more difficult than installing subflooring, except that you are working up in the air and on a slope. Full panels of plywood are staggered in the same manner as the floor sheathing. Do a sketch of the layout first to avoid waste (illus. 25).

Start panel installation at any corner of the roof. It is usually suggested that you use 6d common smooth nails. They can be spaced 6 inches on center along the ends of panels and 12 inches on center at intermediate supports. As with subflooring, leave a space of about 1/16 inch at panel ends and 1/8 inch at edge joints for expansion.

Your plans will show either open soffits or closed soffits. If you have open soffits, you will have to be careful to use nails that are short enough on the

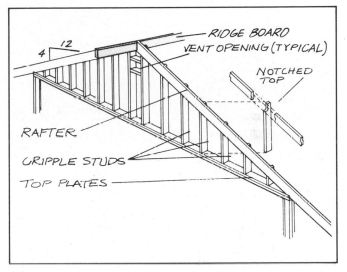

24. *Frame out the roof end and create a vent opening if necessary. Drawing courtesy of American Plywood Association*

roofing material so that you do not puncture the plywood and leave nail ends showing (illus. 26).

Roof Covering

On a pitched roof, one of the most economical roof coverings is asphalt shingles. Today, these shingles come in a great variety of colors and textures so finding one to fit your design scheme is easy. Of course, you can always use cedar shingles. Although quite beautiful, they are also very expensive.

To begin, erect a type of scaffolding so that you can bring shingles to the roof area and also have room to work. A local hardware store or equipment rental firm may be able to help you with this.

The roof must be free of all debris and absolutely dry. Check and correct any irregularities.

An underlayment of roofing felt is stapled to the sheathing. Typically, the felt comes in rolls about 3 feet wide. Roll one course out and secure. Roll the next course out and lap it at least 4 inches over the previous course.

Areas such as the peak of the roof should be flashed to prevent moisture penetration. Flashing is a strip of noncorrosive metal which overlaps the peak. On more complicated roofs where there are a number of peaks and valleys, extensive flashing is needed.

Before you actually start to shingle, distribute bundles along the roof. Shingles come in bundles, and three bundles equal a square which covers 100 square feet of roof area.

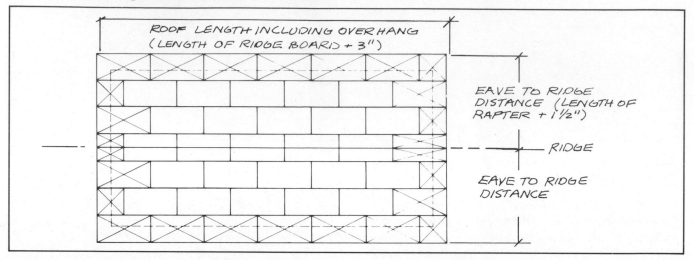

25. Make a layout of the sheathing positioning to cut down on waste. Drawing courtesy of American Plywood Association

26. Shown here are details for open and closed soffits. Drawing courtesy of American Plywood Association

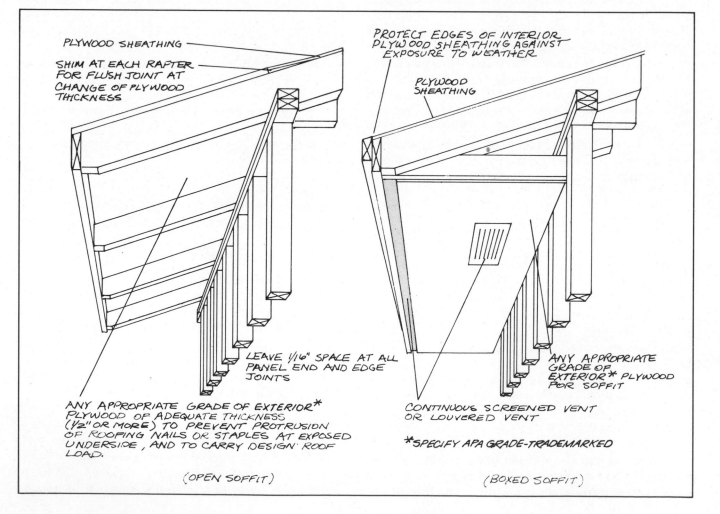

Shingling begins at the eaves, working up toward the ridge. A row of shingles is applied, the next row overlaps, and so on. Carefully measure up 12 inches from the eave and snap a chalk line along the roof to act as a guide for applying shingles in a straight line.

Nail the starter course. Use hot-dipped galvanized nails, two for each tab or six nails for each shingle. Once the starter course is in place, again snap a chalk line and apply the second course. You should have 7 inches of overlap, so take that into account when positioning the chalk line (illus. 27).

This is a quick, basic course in roofing. For special and more difficult situations, it is safest to consult a professional. See *Homeowner's Guide to Roofing and Siding* for more information.

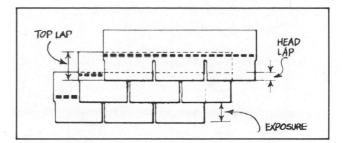

27. *Asphalt shingle terminology includes the roofing terms indicated in this drawing. Whether a homeowner contracts the work or does it himself, he should acquaint himself with the basic fundamentals of the work so that he can talk intelligently with contractors or suppliers. Drawing courtesy of the Asphalt Roofing Manufacturers*

A Weathertight House

Once the roof shingles are in place, windows and doors should be added to make the house weathertight and secure. The best bet for the do-it-yourselfer is to purchase prehung windows and doors. Typically, they are fitted into the rough opening created during framing. Blocks of wood are placed around the unit which is then leveled and secured. Relatively inexpensive interior doors also come this way and can be quickly installed.

Sliding glass doors usually come in a package and must be assembled. Units can be assembled on the flat and then tipped into place. They are usually bolted to the structure.

Mechanical and Electrical When the house is weathertight, rough mechanical and rough electrical work can be done. In virtually every area of the country, skilled and licensed contractors are required. These jobs must be completed before the walls can be closed up. For a better understanding of these systems see *Homeowner's Guide to Plumbing* and *Homeowner's Guide to Electrical Wiring*.

Wall Covering

The two basic types of wall covering or wall finish are plaster and gypsum board. Generally, wallboard is easier to install. It comes in 4 x 8, 4 x 10, and 4 x 12 sheets.

If you have never applied wallboard before, you might want to subcontract it out. Although the actual installation is not difficult, taping and spackling the joints is difficult for a beginner.

Wallboard is installed so that the 8 foot side of the 4 x 8 sheet is laid on the horizontal. Begin the first panel at a corner of the room. Nail through the wallboard directly into the stud about 4 to 6 inches on center. Give the nail an extra hit so that its head is below the surface of the board.

The panels can only butt over a stud, never in between. Once the first panel is in place, the second can be added alongside it. Panels must always be staggered. Cut the 4 x 8 panel so that you have two sections of 4 x 4. Tip it into place and nail.

Watch out for electrical and mechanical fixtures in the wall. You will have to make cuts in the wallboard for these fixtures as you proceed.

Basically, wallboard is applied to ceilings in the same way. Wallboard must run perpendicular to ceiling joists. Each new course must be staggered, and panels must butt only over a joist.

When the wallboard is in place, metal angles are nailed into all corners. Tape and spackle can be applied now.

Spackle can not be applied in cold weather. Where

Gypsum wallboard can be cut with a utility knife, using a 4-foot T-square to assure an accurate cut.

In horizontal wall application of gypsum wallboard, the top panel should be installed first, pushing it flush to the ceiling.

Paper tape and joint compound are applied with a wide knife.

two panels butt, spackle is applied in a long strip. The tape is worked into it with a spackling knife. Spackle is then applied over the tape and pressed firmly over the joint. Excess spackle should be removed. Spackle is again applied after the first coating dries. Repeat this procedure two or three times, until the joint is as smooth as the rest of the wall.

Flooring

There are a variety of flooring types you can select for your house. The installation of oak wood flooring requires a skilled craftsman and special tools. If you are going to finish your own flooring, try tile or carpeting. Unlike oak floor, tile and carpeting require another layer of plywood to make the floor firm before these finishing materials can be applied.

Finishing the House

There are a great many other things you will wish to do to your new vacation home. Some of these things you can do yourself, others you will need help with. Installing appliances and fixtures is better left to a professional while painting or staining your new siding can be completed yourself.

Before you undertake such a project, it is suggested that you do more reading on the subject and fully understand your house plans. See *How to Build Your Own Home.*

If you are going to build a vacation house yourself, keep the overall project small and keep the design simple. Have someone available in the area who is knowledgeable in construction to help you if you have need of it.

A 4-inch joint-finishing knife is used to hide nails below the wallboard surface. Photos courtesy of Georgia-Pacific Corporation

Building a Log Cabin

Log cabins have been a sturdy tradition since the colonists began settling America. In recent years there has been a revival in this mode of building. For some people it's probably a genuine pioneering spirit, but for most others the idea is grounded in economics: a log cabin is relatively inexpensive, is possible for a do-it-yourselfer to construct, and, when compared to other types of vacation homes, is very energy efficient. Depending on wood species, log homes are very durable. Cedar is most popular. Log cabin kit-packaging has become a large business in recent years, with suppliers in virtually every part of the country. Another indication that this form of building is becoming popular is frequent long waiting lists, increasing the time between order and delivery.

It is possible to build a log cabin from scratch. All you need is a good site with plenty of mature trees. You can fell as many trees as you need, strip the bark, and let the logs cure. You don't want to remove all the trees on your site. Your choice of woods will also be limited to the species of trees that grow on the property. Softwoods (cedar, pine, spruce, fir) are lighter and easier to work with than hardwoods.

Softwoods are also better insulators. Please refer to the wood chart to determine the type of wood best suited to your particular needs. If you cut your own trees, they must be dragged from the forest to the building site. The logs can then be notched and hoisted into place.

Log cabins can be built from scratch, but unless you have plenty of time, physical strength, and the willingness and ability to labor long and hard, kits may be easier. If you have a limited source of trees on your property, a kit is recommended, if not essential. Factory-milled logs, although convenient and time-saving, are more expensive.

The chief advantages of building with precut kits are savings in time and labor plus better quality construction than might be obtained with from-scratch building projects. Two men with average construction skills plus a couple of laborers should be able to erect a shell of a medium-sized kit log home in four days. Depending on the design and skill of the builder, it takes about one man-hour per square foot of floor area to perform the basic shell construction in a kit-built home. Another advantage of kits is their mobility: log home builders no

PROPERTIES OF WOODS

Species	Strength	Weight	Working Ease	Decay Resistance	Shrink/ Cracking
Ash, white	high	heavy	hard	low	high
Beech	high	heavy	hard	low	high
Birch	high	heavy	hard	low	high
Cedar, red	medium	medium	easy	very high	low
Cedar, white	medium	medium	easy	very high	very low
Cottonwood	low	light	medium	low	med.-high
Cypress	medium	medium	medium	very high	low
Fir, Douglas	medium	medium	medium	medium	medium
Gum	medium	medium	medium	medium	medium
Hickory	high	heavy	hard	very low	high
Maple, hard	high	heavy	hard	low	high
Oak, red	high	heavy	hard	low	med.-high
Oak, white	high	heavy	hard	medium	med.-low
Pine, eastern white	low	light	easy	medium	low
Pine, ponderosa	low	light	easy	low	med.-low
Pine, yellow	high	medium	medium	medium	med.-high
Spruce	low	light	medium	low	medium
Walnut, black	high	heavy	medium	high	medium

NOTE: Some woods that otherwise might be suitable for house logs may be priced out of practical use. Black walnut, for instance, is a strong, decay-resistant wood that is fairly easily worked; however the price for veneer grade walnut can be several dollars per board foot, on the stump. A builder might get more mileage out of this timber if he sold it for cash, then bought logs of other species for his house.

longer are restricted to the species of logs in their area. In many parts of the country, even heavily timbered regions, builders often prefer to use precut logs for durability or appearance.

This freedom of choice can be expensive, however. Despite the claims by some log home manufacturers, building with precut kits is not cheap. Precut homes cost from $3,000 to well over $20,000 for the kit materials alone, depending on the size and style of house built. This initial kit cost represents from one-third to one-fifth of the total completed cost, depending on how much is spent finishing the house, its design, and how much of the work the owner can do himself. If you do virtually all of the work yourself, including much of the wiring, plumbing, and interior finishing, you can expect construction to cost about two and one-half times the basic kit price.

Are You Ready for a Log Cabin?

Although a log cabin can be built by a person with little knowledge of construction, it should not be attempted by a rank amateur. If you want a log cabin but don't know your way around a construction site, it might be advisable to purchase materials from one of the factories which can also supply a construction foreman for an additional cost. This individual will supervise the construction at your site. The only talents you and your helpers need are strong backs and a willingness to follow directions.

Major log home companies or their dealers often provide a certain level of technical assistance to builders. These services range from planning and design help in the initial stages, to on-site assistance to get the owner-builder started. Generally, these services are included in the price of the kit package.

Precut log home companies provide complete blueprints and detailed architectural plans along with their kits, and many provide materials lists and specifications for building components not supplied with the kit package. Some companies also publish excellent step-by-step construction guides.

Before the kit arrives, a foundation must be constructed. This can be a full foundation, slab-on-grade, piers, or crawl space. Although you can select the type of foundation you want, the factory may suggest a foundation best suited to the kit. The foundation must exactly fit the log kit; therefore, be sure you know both the inside and outside dimensions. Have a thorough discussion with the factory representative, and make sure you understand exactly what is called for. Constructing a good foundation was covered in the preceding chapter.

You will also need a solid road in place so that the big flatbed trailer carrying your future house can get to the foundation site. If the truck can not negotiate your road, the kit will be dropped off at the closest possible point, meaning that you or helpers will have to drag thousands of pounds of logs to the site—a back-breaking and costly job.

Selecting a Supplier

Your choice of a log home kit will be limited in part by the number of factories which deliver in your area. Delivery charges are based on mileage. Depending on the distance from the factory the kit must be delivered, the cost could exceed the total kit cost. Once you have determined which companies will deliver to your area, you can select a kit. Most factories sell a variety of kits. There are many low-cost kits available which are units with only one or two rooms. Factories also sell far larger units with two stories and three or four bedrooms. If you want a low-cost vacation home, stick to the smaller units. They will be less expensive and far easier to build.

Visit the factory or a local dealer and walk through the model log homes to be sure of the size you need. Ask the factory representative if the models are designed so that you can start with a small unit and add to it in a few years. Such models are available.

Several factories have complete engineering and design departments to adapt precut kits to a customer's specifications.

Some companies will make preliminary, scaled floor-plan drawings free of charge, with no obligation to purchase a kit. Most, however, require a signed purchase order and a minimum deposit before any plans are drawn.

With costs of labor and materials steadily increasing, price lists usually carry the warning that prices are subject to change without notice. A purchase contract "locks in" the agreed-upon price, even if costs go up before delivery is made.

Suppliers have different payment schedules. Most factories require a down payment of 25 percent when the purchase order is signed with the remainder on or before the date the first load is delivered to the building site.

Kit packages are commonly shipped on flatbed, semitrailer trucks, either company-owned trucks or by common carrier. In cases where shipment is most economical by rail, that form of transportation is used.

Final payment for the kit is usually made when the materials are delivered. Interstate Commerce Commission rules require a COD charge to be added to the cost of the kit and transportation. A

Log cabins fit in well with vacation or leisure-home life styles. This home is a chalet with two bathrooms and three bedrooms.

customer can avoid this extra charge by making the final kit payment a couple of weeks ahead of shipment. As a result, the only payment necessary when the kit is delivered will be the shipping bill. Larger homes will require two trucks, sometimes three, to haul the material.

If you are going to need a mortgage to finance this project, make sure the bank understands that you need most of the money before construction. Some banks are reluctant to lend money before the house is built. Most prefer that payments be made at each stage; after the foundation is complete, after framing is finished, and so on. Financial details should be worked out long before the logs arrive. Have delivery payment specified in the contract.

Receiving a Delivery

We will assume that all the financial arrangements were set, foundation and platform have been installed, access road constructed, and you are waiting for delivery of your new log home. Try to have three or more people there with you. Truck drivers only deliver the kit; they do not help with the unloading. If you do not have enough help, the driver will drop the logs on the ground where they can be chipped, cracked, and soiled.

When the truck arrives, all breakable items should be stored away from the surrounding area in a secure and relatively dry place. The logs and other items should be placed on a carpet of hay on the ground. Logs and other structural members should

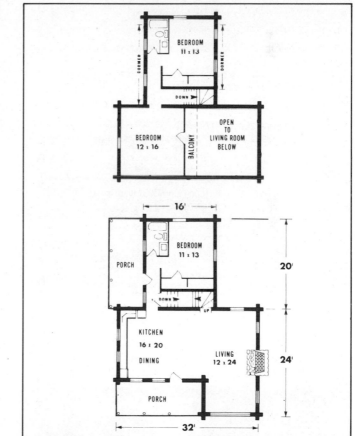

be separated and stored according to length and function.

Once you take delivery of the log kit, you have thousands of dollars worth of building materials on an unprotected site. You must find your own solution to protecting these materials. The best solution is to get the house nailed into place as soon as possi-

Log homes come in many different styles and are popular in both rural and suburban areas. Photos courtesy of Vermont Log Buildings, Inc.

A road for access to your site is essential. Be sure to have plenty of help at the construction site when your new log home arrives. Photo courtesy of Vermont Log Buildings, Inc.

As logs and other materials are unloaded, they should be separated. Breakables should be put in a safe, dry spot; logs should be sorted by length and use. Photo courtesy of Vermont Log Buildings, Inc.

ble. Your vacation homesite may seem like the most tranquil, secluded spot in the world, but building materials have a habit of disappearing when left out in the open.

When you purchase the kit, the salesperson can probably estimate how long it will take to build it. Although one person can build a small cabin, in reality two or more people can put up a cabin much faster. If you can afford it, or manage it, four workers are ideal.

Tools Required

General residential carpentry tools are needed for construction of a log house. These include:

- two 6 or 8 lb. sledge hammers—for driving spikes;
- 3-foot pinch bar—for moving logs and for removing spikes that have been driven in the wrong spots;
- 3-foot level—to check plumb of doors and windows;
- staple gun or staple hammer—for tacking gasket material;
- ratchet winch and 30 feet of rope—for pulling logs tightly together;
- 50- or 100-foot steel tape—for checking various building dimensions;
- a carpenter's claw hammer—for nailing decking, window frames, etc.;
- handsaw or power saw—for cutting joists, plywood, etc.;
- framing square—for checking square on doors and windows;
- wide-bladed wood chisel—for cleaning mortise cuts;
- drawknife—for trimming logs and removing bark;
- chalk line—for snapping lines to align building members.

Other equipment and materials that may come in handy include:

- small chain saw for cutting larger fireplace

Each log in a home kit is coded to identify its position in the building. Photo courtesy of Authentic Homes Corp.

opening
- caulking gun
- wire cutters
- electric drill and bits
- flashing for termite shield
- tin snips to cut flashing
- supply of 2 x 4s and 1 x 4s for bracing and framing
- nails (a good supply of 16d and 8d commons, plus a few 20d and 40d)
- conduit (if needed for electric runs)
- hacksaw to cut the conduit
- dunnage or scrap lumber (such as 4 x 4s) to support the logs off the ground

A chain saw, incidentally, is an excellent investment for virtually every vacation home builder and owner. You can use it while building the house and for cutting firewood later. The log kit supplier will notify you if any exotic tools are required.

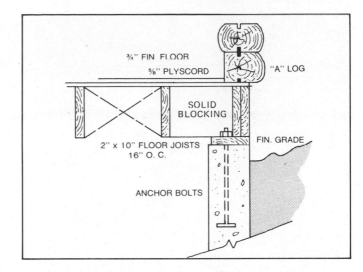

Construction details vary with kit homes. Most use a box sill method for subfloor framing. Photo courtesy of Vermont Log Buildings, Inc.

Materials

Besides the log home kit, you will need other building materials for the foundation, the bottom platform, and the roof. Check these extra requirements with the log home factory. The salesperson can usually tell you what extra materials you need to buy. Before you settle on a kit, add the cost of the kit and everything else you will need to buy. That will be the actual cost of the home. Once you get this real cost, you might want to compare it to the cost of a conventionally constructed home to see if it is indeed cheaper.

Assembly

Assembling the kit is not difficult, especially if you have enough helpers and a factory representative on hand. Suppliers go to the extreme in labeling each piece of the kit. Although they are not foolproof, if you carefully follow instructions, you should not have problems.

The first step is to check level and squareness of the foundation and to construct a bottom platform. Kit manufacturers include in-depth instructions on how to construct the bottom platform.

Because each supplier has its own set of instructions, we will only deal in general terms on how a log home kit is constructed.

The First Course Before the logs are set in place, you have to install a sill seal. This can be stapled into place. With all logs previously separated into piles, you can begin to lay out the first course. It is extremely important to get this first row installed properly because every row you install after it will be influenced by it.

Begin with the front wall of the house and align

all the logs. When they are in line with the foundation, fasten them in place. Usually spikes at the ends of logs are not driven until the adjoining wall log is in place.

With front wall logs in place, sidewall logs can be added. The sidewalls must sit at right angles to the first wall. An easy way to assure square walls is to use the 3, 4, 5 triangle concept mentioned earlier.

Doors Doors must be installed next. Read over descriptions of these units to ensure proper installation.

The doors and frames are installed, leveled, and braced. Bracing is usually nothing more than temporarily nailing a 2 x 4 from the door frame to the platform. These braces will hold the doors in place while you position the logs around them.

Windows The courses of logs are set and secured according to the manufacturer's instructions. Following the instructions closely, you will come to the course where you must add windows. Generally, all the windows are on the same level or course and are added at the same time. Windows and frames are carefully placed on top of the course of logs and then braced. It is not advisable to brace from the side but rather from the top of the frame to the platforms.

Windows must be treated with care. The windows should be kept in a closed position to maintain square. Logs should never be forced in around them. If force is necessary, something is out of square.

Final Courses With windows in place, log courses around windows are filled in, and longer logs are added over windows. To prevent a possible twisting of the structure as it is tightened and secured, alternate the tightening of each course; that is, one time tighten it clockwise and on the next course tighten it counterclockwise.

A few log home companies use round logs, joined on facing surfaces with tongue-and-grooves or splines. Photo courtesy of Beaver Log Homes

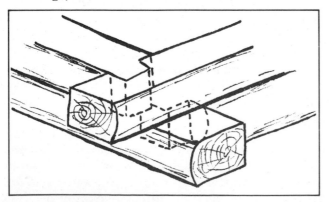

As can be seen in this log cabin during construction, these homes are not difficult to build; however, instructions must be closely followed. Photo courtesy of Vermont Log Buildings, Inc.

The Ward Cabin Company uses squared, white cedar logs with interlocking corners.

Installation Details

Joists You might want to read again the section on joist installation in the previous chapter to understand just what is involved.

With log homes, you will set a final course and then install ceiling joists. Before installing joists, make sure all dimensions are relatively square. If something is out of square, you can use a come-along to help straighten it out. A come-along is a rachet winch which is attached to one section of the house by use of a metal cable. A longer cable is fastened to another section. Using a lever and pulleys within the winch, the longer cable is slowly drawn toward the mechanism. If the framing is slightly out of square, the come-along could be attached to diagonal corners of the framing, and the structure could be slightly drawn in to achieve square. This is a powerful little device. If you have not had experience using it, find someone who knows how to use it properly.

The center girder, if called for in the plans, must be installed first. The girder runs at right angles to the ceiling joists and supports them at midspan. The girder may need the support of a stud wall or several 4 x 4 posts beneath it.

With the girder firmly in place, the joists can now be added. This process requires several people due to the weight of the logs. The joists should, however, slip easily into grooves. With joists in place (but not spiked or nailed), check for level. You will find that some joists are not level. Most kit manufacturers supply shims which are driven into the grooves at the end of the joist to help level it. With all joists level and the building in square, you can secure these supports in place.

When joists are secured, lay down several sheets of plywood and toenail them into the floor joists to create a safe platform on which to work.

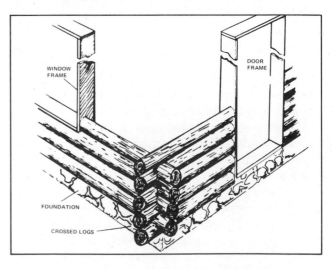

The Wilderness Log Home utilizes round logs with insulation and caulking between logs.

Any builder with moderate construction skill can build his own kit log home. Blueprints and technical instructions provide step-by-step guides. Photo courtesy of Vermont Log Buildings

Porch Logs Depending on design, porch logs are usually installed at this point. Some designs have full porches with roofs while others have decks. Porch logs installed on the vertical can be firmly held with braces.

Gable Ends Gable end logs are then installed according to your plans. Gable ends should never be left unbraced. They present an easy target for the wind and could either be blown over or knocked out of alignment.

Roof Rafters Before roof rafters are set in place, the dimensions of the building should again be checked. Once you are sure of the dimensions, toe-nail in two rafters at a gable end and install the ridgeboard. The ridgeboard should be braced with 2 x 4s to hold it level. With most kits, the ridgeboard comes in several sections. Butt joints should only butt where two roof rafters join, never in the middle of a span.

The rafters should then be installed according to your plans. Start at one end of the house and continue to the other. Spike and secure rafters. Remember, any time you temporarily secure any support member, be sure to go back and fasten it permanently. Often in the excitement of building, something will be secured only temporarily. This

Interiors of log cabins are very attractive, especially when rough beams are left exposed. Photo courtesy of Vermont Log Buildings, Inc.

could cause major problems after construction is complete.

Roofing The roof of a log cabin or log home is finished similarly to the roof described in the previous chapter, or according to the log kit plan. Typically, this means that some form of exterior sheathing is applied, and then asphalt shingles or wood shakes are applied. Wood shakes make an outstanding roofing material if you can afford the additional expense. This material weathers to an attractive color that complements the wood of the cabin.

Filling In Insulation must be applied under the roof, and a vapor barrier must be installed. Insulation can also be used around windows and doors. The unit must also be caulked according to the kit manufacturer's instructions.

Usually, log home kit builders will leave the logs exposed on the interior. Wood, however, is not as efficient an insulation material as standard insulation. If you plan to use the cabin during winter in cold climates, consider using insulation on inside walls and finishing the wall off in either wallboard or interior paneling.

The Interior

Once the unit is weathertight, you can begin finishing off the interior. Many log home builders hurry to finish the exterior to make the structure secure and weathertight and then finish the inside at a leisurely pace.

In order of importance, the interior partitions should be completed as soon as possible. Plumbing and electrical installation should be completed, and then kitchen and bathroom fixtures installed along with the heating unit.

This is a quick explanation of how to build a log home from a kit. The construction process is not difficult if you follow plans and make sure that every course is square and level before moving to the next one.

Building from Scratch Basics

With the foundation poured and the logs cut, seasoned, and squared, you are ready to begin building your home. Ideally, you will have thought through each stage of construction, decided what kind of corner notch to make, how you will tie in floor joists, what kind of subflooring you will put down, what kind of roof framing and roofing materials you will use, and all the other details before you start building.

Laying Out Floor Joists Since homes are built from the bottom up, we will start with the subflooring system, the part of the house which sits directly on the foundation.

Floor joists form a level framework over which the subfloor is laid. When selecting lumber for joists, figure on a width in inches equal to the span in feet. For example, 2 x 8s can be used for spans up to 8 feet and 2 x 10s for spans up to 10 feet. Joists are laid the short way.

If you plan to use logs for interior partition walls, those partitions must be placed on the floor joists to handle the extra weight. Try to place all partitions over a beam or girder, if possible.

Beams, either of steel or wood, are supported by concrete piers which are poured when the foundation wall is made. The beam, or girder, is laid the

Several methods of supporting floor joists suit log home construction. In this photo, a ledge was built into the foundation wall so that joists are flush with the top of the foundation when placed on the ledge.

long way and supports the inside end of the joists if the span is longer than the standard lengths. Except for long joists mortised into the girder, joists sometimes overlap the top of the girder. Therefore, the girder, or beam, should be placed so the inside end of joists are level with the foundation wall or sill. See the detailed illustrations in the previous chapter for more information on foundations.

Many log home builders and kit log home companies use a standard box sill construction. The first course of logs is laid on the subfloor. This requires fewer logs, since the walls begin on top of the floor. Ribbon or rim joists are usually doubled or tripled to support the extreme weight of log walls.

In this method of construction, a 2 x 6 or 2 x 8 is fitted over the anchor bolts and rests on top of the concrete foundation walls. Joist locations are marked off on the sill, along the longest wall.

A straight 1 x 3 can be marked as a layout template. With a steel square, set off the desired joist interval from one end (usually 16-inch intervals measured from the face of a joist to the corresponding face of the next joist).

Mark the joist thickness at the end of the template as a starting point. Measure 16 inches and square a line across the template with a framing square. Put a check mark inside the line to indicate on which

Subflooring is installed over joists; note the diagonal bridging between joists. If needed, insulation can be placed between joists before the subfloor is nailed down. Photo courtesy of Authentic Homes Corp.

Foundation types vary in home construction. This continuous wall foundation incorporates a ledge for floor joists, which will be level with the top of the foundation. Photo courtesy of Authentic Homes Corp.

side of the lines the joists will go. Repeat at 16-inch intervals.

Place the marked template against the sill, its end flush with the sill's end. Transfer the lines and check marks to the sill. In marking the opposite sill, be sure to start at the same end. Also mark joist locations on the girder plate.

Joists should be doubled under partition walls that run parallel to them for extra support. Spike the joists together when doubling or tripling joists, or insert blocking between them to leave space for any planned wiring or pipes. Disregard the additional joists when measuring joist spacing.

Head off joists with double headers and trimmers around such openings as the fireplace and basement stairways as previously described.

When the joists are in place, aligned, leveled, and nailed, you can fit bridging down the center of each joist span. Bridging helps keep joists vertical and transfers loads to more than one joist. You can use predrilled metal bridging or cut wooden bridging from 1 x 3 stock. The metal bridging can be installed more quickly.

Subfloor The subfloor material is generally nailed to the joists. The first course of logs is either spiked to the floor framing or bolted to the long bolts embedded in the foundation wall. The bolts should

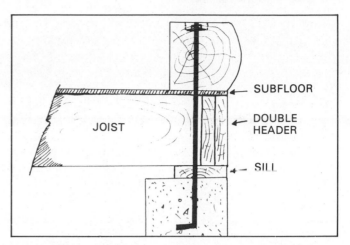

Solidly anchored, logs, subfloor, and sill are bolted to the foundation wall.

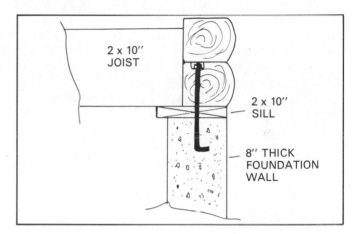

In this drawing, both the sill and first course of logs are bolted to the foundation wall.

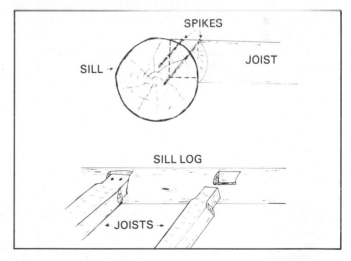

Where log floor joists are used, the ends of the joists usually are mortised into the sill log and spiked to it.

extend through the sill and first course of logs.

Other suitable methods of floor framing can be used. Some builders form a sort of brick ledge in the foundation or basement wall to support the joists level with the top of the foundation wall. Others use

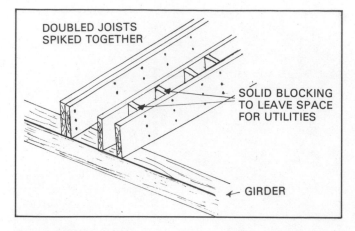

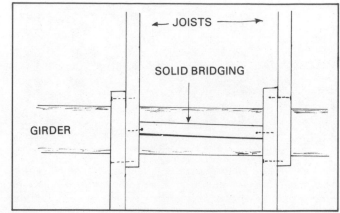

Doubled joists which run under partitions or bear other heavy floor loads should be spiked together. Where plumbing or other utilities will run between doubled joists, separate them with solid blocks of the same joist lumber.

Solid bridging, cut from the same material as joists, can be used to brace joists so they will not twist out of vertical while under loads. Depending on the construction, this solid bridging can be used to brace joists over the girder, or it may be installed at the ends of joists along the foundation wall, or both.

a wider sill and butt the floor joists against the first row of logs or use metal joist hangers to fasten joist ends directly to the bottom log.

Joists made of logs are usually mortised or notched into the bottom log and spiked securely. The girder is level with the bottom or sill log in the wall and is also notched to take the inside ends of joists.

For all methods of floor framing, use joists of heavy material and space them to give sufficient support to the entire floor load. Once the house is built, it is difficult and expensive to reinforce the floor framing.

Raising the Walls At this point, you are ready to begin building the log walls. Here are some preconstruction procedures that will make the actual work easier.

To some extent, you may be able to do part of the work on a prefab basis depending on your schedule and design. You can preassemble door and window frames and headers. You can build gusseted roof trusses or precut rafters and tie beams. You can even cut logs to length ahead of time, if your design allows that much precision. The more work you get done prior to actual construction, the faster things will go at the site.

Your prebuilding work will need to be done with precision. Cutting a full set of rafters to the wrong length would waste time and material.

Solid-log construction is not instant housing, even if you prefabricate several components ahead of time. You are building a home from the ground up, and it takes time to cut and fit 300- to 600-pound logs into a strong, tight wall.

Unless you have four or five helpers, you'll need some kind of mechanical equipment to hoist the logs into position, such as a tractor with a front-end loader, gin-pole type of boom and winch, and an

inclined ramp with block-and-tackle. What you use will depend on the equipment available in the area. Make sure it is heavy enough to do the job safely.

Although not the fastest method of hoisting logs, tripod swinging booms with block-and-tackle can be effective. You can set up a boom at each wall and move the tackle from one to another.

Once the walls are to the height that window frames are to be set in, you can move several logs over the low wall onto the floor deck. They will be easier to put up from the level surface inside if the floor is braced well enough to take the load.

Builders who use logs sawed flat on three sides often build a form of plumbed uprights around the perimeter of the house. A chalk line is snapped along the inside wall line, and 2 x 4 and 2 x 6 uprights are plumbed and braced flush with this line. Another method is to lay the first row in position, and the uprights are plumbed against the inside of these logs. Each course of logs is placed into position snugly against these uprights and spiked or bolted together. This is an easy way to get a straight, plumb inside wall. Each course must be leveled.

Some builders do not spike or pin logs together, depending on the weight of the logs, to hold them in place. It is good policy, however, to fasten logs together every 8 feet or so around the wall and at corners and wall openings.

One method of fastening is to spike each row of logs to the one below it with 10-inch galvanized spikes. Another is to drill pilot holes into two or three logs at a time, then drive sections of half-inch-diameter reinforcing bar into the holes.

One precut log manufacturer uses 8-foot-long threaded rods to bolt the entire wall together at intervals of 8 feet.

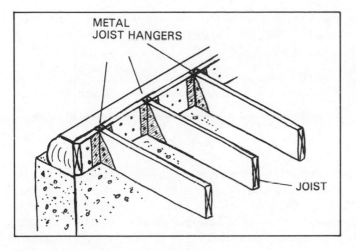

Metal joist hangers can be used.

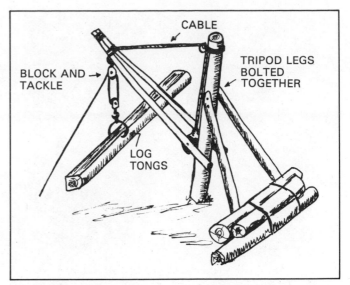

A tripod boom with block-and-tackle can be used to hoist heavy logs into position.

If you plan to do much hole-boring, use a heavy duty electric drill fitted with a long drilling bit. If electrical service is not available at the site, you can rent a gasoline-engine auger. The alternative is to bore holes with a hand auger or brace-and-bit.

Squared logs can be laid one directly on the other with a gasket of urethane or other insulating material between them. Joists between logs should be caulked to make the wall watertight.

Corners With round logs, the simplest corner notch is a round notch cut to half the diameter of the log. This notch is made only in the top log. Cut the notches to fit snugly over the logs below and caulk all joints well.

For squared logs, the fastest corner to make is the abutted "cob" corner. Alternate courses of logs extend about 1 foot past the corner of the house. The key to making good cob corners is to measure logs carefully and cut the ends of abutting logs squarely.

An even stronger corner for squared logs can be made by notching the extended log about 2 inches and recessing the abutting log into the notch. It takes more time, but the interlocking logs make a sturdier structure.

Simple or compound dovetail notches make strong corners. The drawback with dovetail corners is that they have a greater horizontal-cut surface area where moisture can collect.

If possible, logs should be used without splices. Use the longest logs in the walls below windows.

Doors and windows are commonly set in place as the walls go up, and frames serve as stiffeners to insure a plumb opening. In this home, doors and windows are keyed to logs with hardboard splines. Photo courtesy of Vermont Log Buildings, Inc.

Doubled dovetail corner notches are somewhat compli-cated to fit, but when properly cut and joined they make a strong, interlocking corner. This cabin of squared pine logs is more than 100 years old.

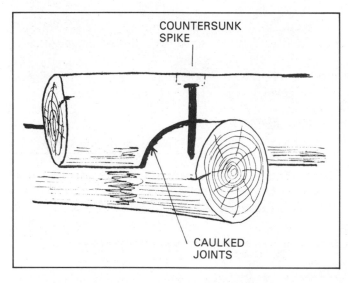

Round corner notches should be made with the notch cut into the upper log so that moisture can not collect in the notch.

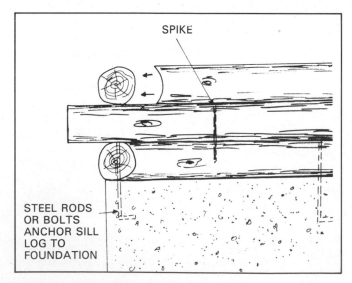

Save shorter logs to fit between windows, doors, and corners. A little planning can save a lot of cutting and splicing.

Wall Openings Depending on the species of wood and how well-seasoned the logs are, log walls may shrink and settle. If the settling is minor, as with seasoned softwood logs, the only result may be some fallen caulking or chinking that needs to be replaced.

If you expect logs to settle quite a bit as they dry in the walls, allow space for movement over doors, windows, and other openings. Fasten door and window frames to lower logs at the bottom and on both sides. Leave 1 inch or so of space between the top of the frame and the log above. Nail a drip-cap and inside trim only to this log, not to the window frame. The log can now move without binding the door or window. The space between the top of the frame opening and the log above should be filled with insulation.

Another solution to this potential problem is to cut a "key" way into the ends of logs that butt against the window frame to fit around a 1 x 1 or 1 x 2 spline nailed to the sides of the window frame. As logs settle and shrink, they can move without bind-ing and warping the window. This can also be done with doors and other openings.

Lay logs to the top of windows and doors. On the next row of logs, the one that will form the lintel over these openings, attach metal flashing or drip-caps over doors and windows. Measure off on the log and attach the flashing to what will be the bottom side of the log when it is in place. Use cop-per, galvanized metal, or aluminum metal for the flashing.

If the roof will overhang by 2 feet or more, you can forego the flashing and simply nail a piece of 1 x 2

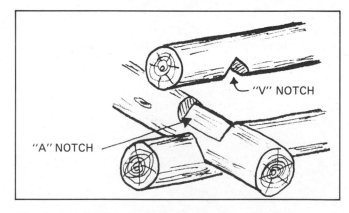

The "A" and "V" mating notchings are most commonly used for round logs but can be adapted to squared logs as well.

Cob corners can be made with round logs, too. The abut-ting log end should be cut to fit the curve of the log it joins.

trim to the top log to cover the space between the window frame and the log above.

Windows Try to plan all windows the same depth. If you use logs of uniform thickness, the windows will sit on the same course of logs and will be covered by the same course of logs. When installing window units, make sure windows are closed and latched and that they are positioned properly. Place the window unit on the log it will rest on, and plumb it by moving the top in or out. Brace it with a length of 1 x 4 or 1 x 6 nailed to the deck. Also, cut a stiffening spreader to nail horizontally across the window frame, about halfway up. Nail the brace to the frame at the top, not the sides. See the next chapter for more information on doors and windows.

One Story or Two? First-floor walls should be at least 7 feet, 4 inches high. An even 8 feet can save a lot of cutting if you are planning to finish interior walls with paneling, Sheetrock, or other wallcovering that comes in standard 8-foot lengths.

When the walls reach the desired ceiling height, you are ready to set second-floor joists, ceiling joists, or tie beams depending on the design of your house. If your log home is a single-story structure, you will frame the roof at this point. If a loft or second floor is planned, continue to lay courses of logs above the ceiling joists to make a full wall or stub wall, then frame the roof. A short sidewall provides more usable space in the attic.

Log joists are usually notched or mortised into the log wall and into a girder supported by posts in the center of the house. The girder should be 8 x 8 or larger. Supporting posts are placed directly over the girder that runs under the subfloor joists. If a partition wall will run under the girder, stud this wall before setting the ceiling joists.

Make mortises at the top of the wall logs and in the girder with a chainsaw and large chisel. Saw notches should be a uniform width and depth. Joists should be shaped with a tenon, or tongue, that fits tightly into the mortise.

It is imperative that the tops of the joists and girder create a level surface on which to nail ceiling material or subflooring. Use shims to level the joists. If your design calls for tie beams and purlins (horizontal roof framing members) that will be exposed, a level surface of tie beams is not critical.

At this point, frame out porches and decks if these features are to be included, then continue laying the walls or frame the roof as your design requires. However, if you are building gable ends of logs, install them now. It is easier to build the gables on a level surface of plywood placed over joists and stand the whole assembly in place. Nail a pair of rafters together to use as a pattern to assemble the

Joists and girder should make a level surface to install subflooring or ceiling materials. In this construction, joists also tie walls together.

Where partition walls are placed under second-floor girders, they should be erected before joists are attached to the girder.

gable. Plumb and brace the gables in place until the roof decking has been nailed in place.

Framing the Roof With a one-story dwelling, you have several framing options:

1. use open tie beams and log rafters;
2. use gusseted trusses and conventional ceiling materials;
3. use conventional rafters and ceiling joists; or,
4. use a girder, log joists, and log rafters, as previously described.

Most log homes are built with gable roofs. These are easy to frame and shed rain well when pitched with a wide enough angle. Other roof styles can be used. Log homes can be built with hip, gambrel, and mansard roofs, but these styles require more complicated framing. Log rafters and joists are aesthetically more suitable for a log home, but a smooth roof line is more difficult to attain. Log rafters are harder to cut to precise framing angles than standard lumber.

Rafters are supported by the top logs in the longer side walls. Many builders nail 2 x 6 plates to the top

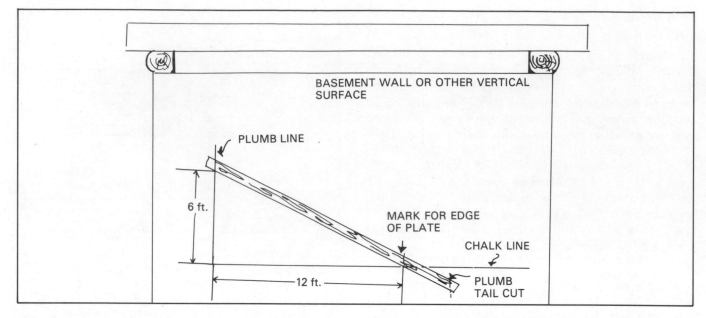

You can use a wall or other flat surface to measure and mark common rafters for gable roofs.

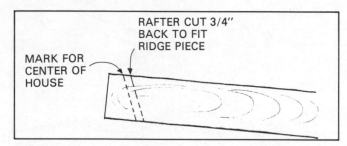

The plumb cut on a rafter should allow for the thickness of the ridge piece.

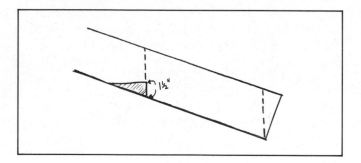

Marking and cutting the bird's mouth notch so it will lie flat on the wall plate.

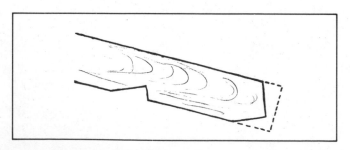

The rafter tail cut to provide a nailing surface for the soffit.

wall logs, then notch rafters to rest squarely on this plate.

The slope, or pitch, of a roof is described in units of rise and run. Rise is the vertical distance from the top of the log wall to the peak of the roof. Run is the horizontal distance from the exterior wall to the center of the house. Pitch is expressed as a number of inches of rise for each 12 inches of run.

The upper rafter ends meet at a ridge piece, usually constructed of a 2 x 8 for conventional rafters. The ridge piece should be the same length as the plate or top log.

The important angle cuts—at the ridge end of the rafter, at the notch where it seats on the plate and the plumb cut at the tail are laid out with a steel framing square. Using a framing square is easier to demonstrate than to explain on paper. You can have an experienced carpenter measure and cut one rafter as a pattern, then use it to cut the rest of the rafters. The measurements marked on the framing square will aid you in determining angles.

This method works for most gable-roof rafters: make a horizontal chalk line across a wall or other smooth vertical surface. If the inside of your log walls are plumb, you can use an end wall. Use your level to make sure the line is perfectly horizontal. Let's assume that you are making rafters to span a 24-foot-wide house with a 6-in-12 roof. Measure off 12 feet along the horizontal line, and make marks that will represent the center of the house and the exterior wall's edge. Using a long level on the center mark, plumb a line at least 6 feet long. Measure 6 feet vertically along this line and make a mark to represent the roof ridge.

Place a piece of rafter stock at an angle so the lower edge of the board barely touches the two marks that represent the roof ridge and wall plate.

The top end of the rafter should extend past the center line by a couple of inches; the bottom end should extend past the mark that describes the plate by enough distance to provide the roof overhang you want.

With someone holding the rafter in this position, transfer the center vertical mark to the rafter, by marking the top and bottom edges of the rafter where they intersect the line. Now, move to the tail of the rafter and, using the framing square against the horizontal chalk line, mark a plumb line on the rafter at the exact point that represents the outside of the wall plate. Measure off the amount of overhang, then make a plumb line to indicate the tail end of the rafter.

The rafter now is marked with three parallel lines, indicating the center of the house, the edge of the wall plate, and the tail of the rafter, respectively. At the ridge, however, a ridgeboard will be placed between the rafters from opposite walls. The rafters will not extend quite to the center of the house at the ridge, but will need to be cut to a point that is half the thickness of the ridge piece. If you are using standard dimensional lumber, say a 2 x 8, for the ridge piece, mark the center plumb cut of the rafter ¾ inch back from the original line. (Nominal 2-inch lumber is actually 1½ inches thick. Half of 1½ is ¾).

At the line which represents the exterior wall, measure and make a mark 1½ inches from the edge of the rafter. The 1½ inches will let the rafter notch equal the thickness of the plate. Place the framing square so that one leg lies along the line representing the wall and the other leg forms a right angle at the mark of 1½ inches. Draw the right angle from the mark toward the ridge end of the rafter. The two lines form a right angle, called the bird's mouth. This notch lets the rafter sit flat on top of the plate with the vertical cut against the exterior side of the plate. The horizontal cut lies flat on top of the plate.

The overhanging end of the rafter can be cut in several ways, depending on how the eaves will be finished. After all lines are marked, cut the rafter or pattern stock. Make the cuts carefully because this will be the pattern for the remaining rafters. Before you cut any rafters, be certain that the walls are perfectly square. If they are not, some rafters may not fit properly.

Measure a distance equal to the ridge height on a straight 2 x 4. Plumb this 2 x 4 in the center of the house. Sit the rafter's bird's mouth in position on the top plate, and line the top of the rafter's ridge end with the mark you made on the 2 x 4. All cuts should fit flush and even.

Log rafters can be exposed in the ceiling. Ceiling

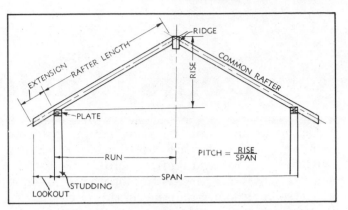

Detail of roof rise, run, and overhang or "lookout."

Log gables must be well braced until roof decking is nailed down.

material is installed over these rafters. A vapor barrier of plastic, aluminum foil, or roofing felt goes on the ceiling, and the roof is blocked out for sheathing with 2 x 4s or 2 x 6s. The space within the blocking is filled with insulation. The roof is then decked with exterior grade plywood. Shingles or other roofing material can be fastened directly to the plywood. This method of roofing increases building costs but combines a well-insulated roof with the rustic appearance of open log rafters.

For a single-story house that will have a finished ceiling, gusseted lightweight trusses can be used. These nailed and glued trusses are a strong, relatively inexpensive way to frame roofs. They can be made up ahead of time. Trusses incorporate both the rafters and collar or tie beams, and if properly designed, can span lengths of 30 feet with no interior load-bearing walls. This allows more flexibility in planning interiors, since partitions need not be placed where they will support ceiling joists.

Doors, Windows, and Decks

Construction details for such features as doors, windows, decks, porches, and stairs are similar, whether you are building from scratch, from precut logs, or rebuilding an existing log structure. These features are grouped together in this chapter.

Stairways should be framed when floors are framed. Make a provision for basement stair openings in the first-floor framing. Main stair openings will be framed into the second-floor system.

Porch and deck framing should be incorporated in the framing of the main floor. These features should be planned along with the basic house design.

Doors and Windows

A tight, weatherproof fit around doors and windows is possible if you install them as an integral part of the wall. There are several ways to do this. You can build roughed-in casings to be fitted in the wall or use complete window-and-sash units and door-and-frame units from lumberyards, log home companies, or other precut home manufacturers. If you are building a precut home, door and window installation is generally an integral part of the design.

If you buy door and window units from a log home company to be installed in a house you are building from scratch, the frame will be fitted with the features that company uses to integrate them into the wall. You will need to shape the abutting ends of logs to accommodate the interlocking feature of the window or door unit.

Rough-in casings should be built of heavier material than you would normally use for frame construction. Two-inch dimensional lumber is best, with the frames squared and braced firmly.

Doors or door frames usually are set directly on the subfloor with the logs butting on both sides of the opening. Windows are set into the wall at a predetermined height. If you are building from a kit, one course of logs will be marked as sills for windows. In construction from scratch, plan to install windows or frames on one course of logs. If your logs are of uneven thickness, some notching and cutting will be necessary to get windowsills at a uniform height.

All logs settle somewhat after being laid. Allow for settling and shrinkage when door and window frames are fastened to the walls. Leave headspace at the top of openings to allow for movement of the logs. Usually, ¾ to 1 inch of space is sufficient, unless the logs are green. The settling allowance gap left between the top of the frame and the log above should be filled with insulation and then covered. If your roof has an overhang of 2 feet or more, you may not need the metal flashing or drip-cap.

When fastening door and window frames to the ends of abutting logs, again take into account the settling of the logs. If the sides of the frames are rigidly attached to the logs, the frame may bind as the logs settle.

One way to allow for this is to make a 1-inch vertical slot in the frame. Drive a nail through the slot near the top, into the end of the log. Do not draw the nail up tight, so the log is free to move downward without putting pressure on the frame. You can attach the frame to several logs for support, using slots each time so logs can move without warping the frames. Do not use rigid insulation in the opening above the window frame. This defeats the purpose of leaving the gap.

You can buy preassembled door-and-frame units for both exterior and interior doors, or you can make your own to suit the overall appearance of your

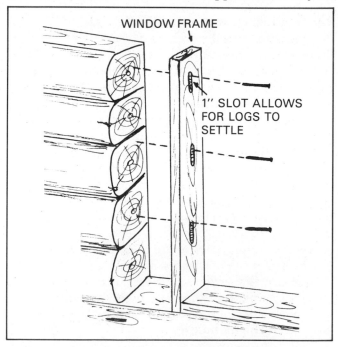

WINDOW FRAME

1″ SLOT ALLOWS FOR LOGS TO SETTLE

To allow for settling of log walls, window frames can be attached to logs with nails driven into a slot in the frame and into the ends of logs. This lets the logs move without warping or binding windows.

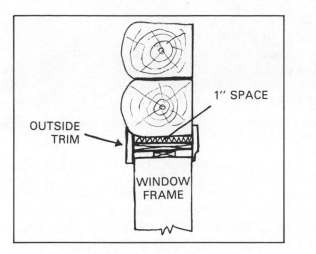

Headspace to allow for log settling should be left above window and door openings. The space can be covered with flashing or trim.

Windows and doors in homes with a wide roof overhang can be finished on the outside with trim, rather than metal flashing. Be sure that trim is nailed only to the log and not to the window or its frame.

home.

Two-inch tongue-and-groove lumber with Z braces will make a rustic-looking door. More rustic but harder to build are doors made of sawmill slabs taken off as logs are sawed. The slabs can be faced for a tight fit and joined together to make a door.

Porches

An attached porch that enhances the house design adds to the overall appearance of a home. The connection of porch framing members to the main house should be by sills, joists, and roof sheathing. The porch rafters, joists, and studs should be securely attached to the house framing.

You can have the porch roof continue the slope of the house roof, or the overhang at the gable can be extended to partially shelter the porch. If more headroom is necessary on the porch, you can make just enough pitch to the porch roof to provide good drainage.

Open porch floors, whether of wood or concrete, should slope away from the house slightly for proper drainage. Floor framing for wood porches should be at least 18 inches above the ground. Wood used for porch floors should have excellent decay resistance and be free of splintering and warping. Species commonly used are cypress, cedar, and redwood. Treat the wood with a preservative where moisture conditions might cause problems.

Supports for enclosed porches usually consist of fully framed stud walls. Studs on 24-inch centers, rather than 16-inch centers as in interior partitions, are ample support for most porches. Double the studs at openings and corners.

In open or partially open porches, solid or built-up poles, posts, or columns should be used. You can use straight, peeled and trimmed poles for open porches.

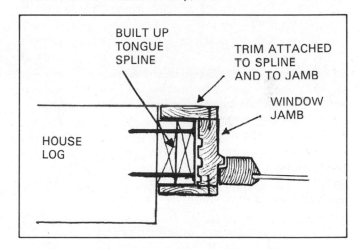

A tongue-type spline can be nailed to log ends, which fits snugly inside trim nailed to each side of the window or door casing. This lets the log move without binding the window.

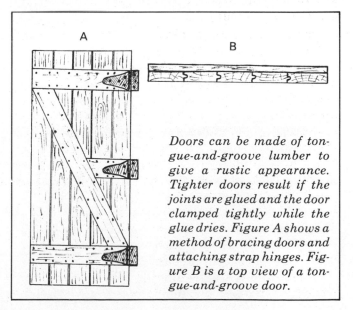

Doors can be made of tongue-and-groove lumber to give a rustic appearance. Tighter doors result if the joints are glued and the door clamped tightly while the glue dries. Figure A shows a method of bracing doors and attaching strap hinges. Figure B is a top view of a tongue-and-groove door.

For a more finished appearance columns can be made of doubled 2 x 4s, which are covered with a 1 x 4 casing on two opposite sides and by 1 x 6s on the other sides.

Slats or grillwork used around an open crawl space under the porch should have a removable section for access to areas where termites may be present. Use a vapor barrier of plastic or roofing felt over the ground under a closed-in porch. A fully enclosed crawl space should be vented.

Decks

Outdoor living is popular with most American families. A wood deck can be a seasonal living room, adding space to a home at modest cost. On hillside sites, a deck can provide a comfortable, accessible outdoor area near the house.

Most decks are attached to the house for access support. As with porches, it is an advantage if the deck design can be incorporated into the design of the house. The deck can be an outdoor extension of a kitchen or dining room, perhaps with access through sliding doors. A high deck above a porch or lower level deck can provide accessible outdoor areas to both floors of a two-story house.

Deck location helps determine the type of deck to build. The location should take into account prevailing winds, as well as sunny and shaded areas during the entire day. A deck facing east will be shaded for evening cookouts and parties. A high-level deck, facing south off an upstairs bedroom, catches late-evening breezes after the sun has set and offers more privacy than exposed low decks.

Some kind of footing should be used to support the posts or poles which transfer deck loads to the ground. A concrete footing is usually most satisfactory. The footing should be placed below the frost line and below any fill dirt. Footings should be a minimum of 12 inches square by 8 inches thick. Where poles or posts will be spaced more than 6 feet apart, a footing 20 inches square by 10 inches thick should be used.

Posts or poles that will support the deck should be located above ground level. A pedestal-type footing which extends at least 6 inches above the ground surface is a good underpinning for most decks. Use nonstaining, rustless fasteners in deck construction. Bolts hold more securely than nails. If posts must be used below grade, they should be pressure-treated with a preservative.

Low wood decks over level ground can be supported on concrete piers or short sections of logs. Soil must slope away from the area under low-level decks to ensure drainage and moisture control.

Decks commonly are floored with spaced planking of decay-resistant, nonsplintering wood. Some are floored with plywood that is sealed with an epoxy coating or other waterproof treatment.

Wood for exposed decks should be pressure-treated with a preservative. Regular applications of a preservative that is brushed or sprayed on can add several years of life to exposed wood. When treating the wood, apply at least two coats to joints and exposed grain ends.

Stairs

Generally, stairways are constructed of materials that complement the overall design and construction of the house. Rough-sawed planks or half-log open stairs lend a rustic appearance, as do railings and balusters made of peeled poles.

Openings in the floor for stairways are framed out during construction of the floor system. The long dimension of the stairway can be either parallel or at right angles to the joists. It is easier to frame a stairway opening with its length parallel to the joists. This is a stronger construction, unless support is placed beneath joists headed at right angles to the run of the stairs. Stairs in split-level houses may have to be framed at right angles to the joists.

Rough openings for the main stairway are usually 10 feet long, depending on the height of the second floor above the main floor. Widths should be 3 feet or more. For basement stairs, a rough opening may be about 9'6" x 32". The width of 32 inches is two joist spaces.

The width of main stairs should be at least 2 feet, 8 inches, inside any handrails. When the stairway runs between two walls, some stairs are designed with a distance of 3 feet, 6 inches between the center lines of sidewalls, resulting in a net width of about 3 feet. Split-level entrance stairs are commonly wider, up to 4 feet or more.

There is a definite relationship between the height of the riser (the vertical step up) and the width (depth) of the tread (the horizontal run) in a stair. Steps with shorter risers usually have deeper, wider treads. If the combination of rise and run is too great, there is strain on the leg muscles. In a stair with a high riser, or vertical distance, the legs are lifting the body's weight more vertically. The feet can not reach a great forward, horizontal distance and still give the legs the leverage they need to pull the body up the stairs.

The reverse is nearly as awkward. Stairs with short risers are an easier climb, but there is a greater tendency to take a longer forward step. If the short riser is accompanied by a narrow tread, the climber kicks his toe against the riser with each step and must make a conscious effort to take small steps. The relation of riser-to-tread area influences walking down stairs as well as climbing.

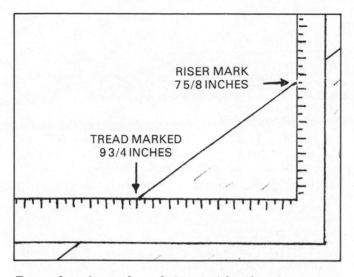

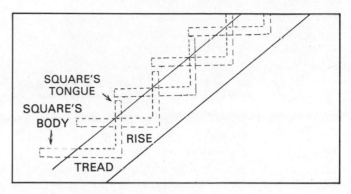

Repeat the marks for each tread and riser, carefully aligning the framing square each time.

To mark stair treads and risers with a framing square, place the square so that the tread width is on the body and the riser height is on the tongue of the square. Mark the 90° angle thus made.

Here is a rule of thumb for figuring the relation between the height of the riser and the width of the tread: the tread width multiplied by the riser height, in inches, should range from 72 to 75. A riser height of 7½ inches should be matched with a net tread width of 10 inches (7½ x 10 = 75).

A riser from 7½ to 8 inches high, with the appropriate tread width, combines both safety and comfort.

To lay out a stair, first measure the total rise (vertical distance) from the finished floor at the foot of the stairs to the finished floor at the top of the stairs. If the finished flooring has not been laid, as is usual when you build the stairway, include the thickness of the floor material that will be used in the measurement.

Since this distance is to be divided into stair risers of equal height, the first computation is to find riser height that divides equally into the total height.

Since the top step will be onto the upstairs floor and the joist or header will be the riser for that step, we can proceed to lay out the other equal steps on the stair carriage or stringer. Figure a tread depth that will make the total rise-and-run for each riser and tread combination fall within the 72 to 75 rule. The total horizontal run (depth from front to back) of the stairway will be the sum of all tread depths.

The stair carriage can be of any material—whole logs, half logs, slabs, or dimension lumber. The procedure for laying out stairs is the same. Round logs are more difficult to mark and cut accurately.

For example, we will show a stair carriage made from 2 x 12 stock. Start at one end of the 2 x 12 and

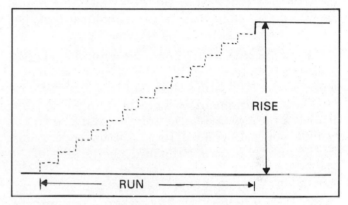

To lay out a stair, start by measuring the total rise. Be sure to allow for the thickness of floor materials to be installed later.

place the tongue of a framing square so that the riser height mark on the square aligns with the edge of the board. Hold the tongue of the square at this point and swivel the square so that the tread width mark on the body of the square also aligns with the edge of the board. Mark the right angle. Repeat this procedure until all risers and treads have been marked.

At the last corner of the last 90° angle marked, which represents the tread at the top of the stairway, mark the end of the stringer so that it will be cut to rest flush against the joist or header. This cut is made at right angles to the mark for the tread, but in the opposite direction than if you were laying out another riser, so that the mark extends to the opposite edge of the stringer.

A horizontal cut will need to be marked at the bottom of the stringer, too, where the piece rests flush against the floor.

By marking carefully, you can use the first stringer (after the treads and risers have been cut out) to mark the cuts for the second stringer.

Kits, Prefabs, and Packages

A simple and quick way to build the vacation home of your dreams is with a kit, prefab, or one of the other packaged homes on the market. Log cabin kits are discussed in the chapter on log homes.

Most of the kits available are of standard materials which are cut, hammered, sawn, and framed in a similar fashion to conventional construction.

Advantages

Why build with a kit? There are a number of good reasons. If you are building the house yourself and want a special feature that requires difficult construction techniques, it will be easier for you to build it with a factory-produced unit. Even if you do not build it yourself, you will have a better understanding of what you are getting and a better idea of the costs.

You may also complement your own do-it-yourself skills with a factory package; there are generally many building options available. The factory crew can do part of the job or all of it. The less they do, the more you save. While you might not want to frame the home yourself, perhaps you would like to finish the interior. There are many other alternatives: you can complete the jobs you feel confident in doing and leave the others to the factory crew.

Another reason to have the exterior shell completed is that you can store other valuable building materials inside. If a house has only been partially constructed, it is vulnerable to theft and vandalism.

Choosing the Right Design

There are many manufacturers of packaged homes on the market today. The largest of these advertise in national and regional magazines. Because of the regional nature of the business, many good factory-home producers can be located through local and regional newspapers. Shipping costs can be quite high, so it is best to find a dealer near the location where you plan to build.

Standard Styles

Manufactured housing consists of precut, panelized, prefab, modular, and mobile home units. Each type offers certain benefits depending on how much you want to do yourself, how fast you want it done, what you want it to look like, and how much you are willing and able to spend.

Precut The precut unit is probably the simplest one that is produced in the factory. The manufacturer precuts and prefits every part of the rough shell or finished house. There are a variety of designs including A-frames, chalets, and contemporaries. This package is an excellent choice for the do-it-yourselfer.

Precut packages come in all price ranges. Large, expensive vacation homes can cost between $30,000 and $50,000, or more. If budget requirements are tight, you can get a precut cabin-style vacation unit for only a few thousand dollars.

Panelized Panelized is another form of manufactured housing used extensively for first- and second-home development, as well as for condominiums and apartments. Wall, roof, and floor sections are put together on high speed machines. Some large lumberyards offer panelized units for sale. Design choices are limited.

Generally, panelized homes can be erected at a much faster pace than precuts. Usually this is not a good do-it-yourself project because panel fastening techniques require special skills and knowledge.

Panelized houses are usually made either in panels or in complete sections, with the plumbing and electrical systems installed in the wall. These units are erected on the building site and are almost immediately ready for occupancy. Only a little money can be saved by making the connections yourself. These are complicated connections better left to an experienced, certified person. In some areas building codes may require that you have professional help.

Prefabs Prefabricated homes offer some savings over conventionally built homes. Although there are a variety of designs available, they are not as extensive as those found with precut or panelized units. Prices range between $15,000 and $20,000, and up. These units originally had a reputation for being produced for the bottom of the housing market, but today their quality has improved. The prime benefit of a prefab is that you can have a fully erected vacation home in a few days or weeks.

Modular Modular homes have become more popular in recent years. These homes are usually larger than prefabs and arrive at the homesite more or less complete. They are less expensive than conventionally built homes.

Because of highway laws, the units must be no

Prefabricated homes, kits, and packages come in many forms, from extremely simple dwellings to high-fashion homes. Here is a precut unit that is fully compatible with the site. Photo courtesy of Acorn Structures

more than 14 feet wide and no more than 64 feet long. Design options are more limited than with other types of manufactured homes. These units are not for the do-it-yourselfer. They are for people who want a vacation home fast and at relatively low cost.

Mobile Homes Mobile homes are a type of modular housing. If you have never thought of a mobile home as a viable vacation home alternative, check into it. Mobile homes are not trailers anymore. There are some designs which appear to be conventional housing.

Mobile homes go into place fast. Although they are limited by highway laws, there are double-wide and triple-wide units available. That is, the house will be delivered at the construction site in two or three units which are then connected together side-by-side to form a big house.

There are a few drawbacks with mobile homes to consider. Some jurisdictions will not permit their use. One other difficulty is that many lending institutions will not approve conventional financing for a mobile home. Instead, they will give you a high-interest short-term financing arrangement.

Selecting the Manufacturer

There are many producers of manufactured homes, and selecting one over another can be confusing. There is a process of elimination to help you narrow the choice. For any particular vacation homesite, there are only a limited number of factories which

can service that location. Some mobile home factories distribute on a national basis, but most deal on a regional basis—perhaps a radius of 500 or 600 miles from the factory. Your choice will be further limited to those units within your price range and design tastes.

Select the house you want from a model, not from literature. If a local dealer does not have the model you are thinking of purchasing, visit the factory where there usually is a larger selection of models available.

Be sure to ask questions. The most important is, how much will the completed house cost. If the factory can not give you a firm price, consult the local contractor in your area who would have the job of assembling it. You may decide on a lovely model home which has a package price of $15,000. Unfortunately, that $15,000 package may cost another $20,000 once it has been delivered and constructed. Before you sign a contract, be certain of the shipping charge.

The delivery date may be another problem. If you order a house on June 15th, do not expect it complete and ready to move into by July 4th. Although the package does not take long to produce or to erect, once the building season begins, there may be a long wait. It is best to order the package during late autumn or winter if you want it ready by spring.

Construction Notes

Site Access One other factor to consider, in addition to design, is the site. Truck access to the site is important. The best kind of site for a factory-built housing unit is one that a truck can drive to directly and can unload material next to the foundation. If you have to haul materials even a short distance, it can greatly add to labor costs.

This is usually not a problem in a developed vacation-home area. If you are ordering a precut home from a manufacturer, the materials can probably be transported to the building site on a jeep or pickup truck. Larger packages, however, are too unwieldly for that.

Interior Finishes For interior finishes, some factory home producers will give you the option of purchasing the materials of your choice for the interior finish at wholesale plus ten percent. Others will rely on their dealers to help you with this. Still others leave the matter of finishing the interior totally up to you. Have this point specified in the contract.

Building Blocks One variation on the concept of designing your own home is literally putting building modules together to create a serviceable home. One New Hampshire manufacturer has come up with a system that comprises three elements: a

12' x 12' module, a 9' x 12' porch that can be enclosed, and a deck. The module is a shell of post and beam framing with a shed roof.

By coupling modules, porches, and decks, you can have a "customized" design for a fairly modest price. One-, two-, and three-bedroom designs can be created, with space ranging from 450 square feet to more than 1,100 square feet. This type of system is geared for amateur builders.

Unusual Shapes For the adventurous, a cabin or an A-frame or a conventional rectangular-shaped home might not be satisfactory. The geodesic dome is one answer; the hexagon is another.

Like the dome design, a hexagon offers great inte-

rior design flexibility because none of the walls are load-bearing. Since these packages are generally fastened together with nuts, bolts, and screws, sections can be added or disassembled for a change in design. Tools needed include: wrench, drill with attachment for self-tapping screws, caulking gun, and ordinary carpenter tools.

Hexagon designs provide more actual living space per square foot than conventional designs. A hexagon-shaped home of 700 square feet features about 30 feet of wall space, for a net living space of 670 square feet. A rectangular-shaped home of 32 feet by 22 feet, or 704 square feet, features 141 square feet of wall space, for a net living space of 563 square feet.

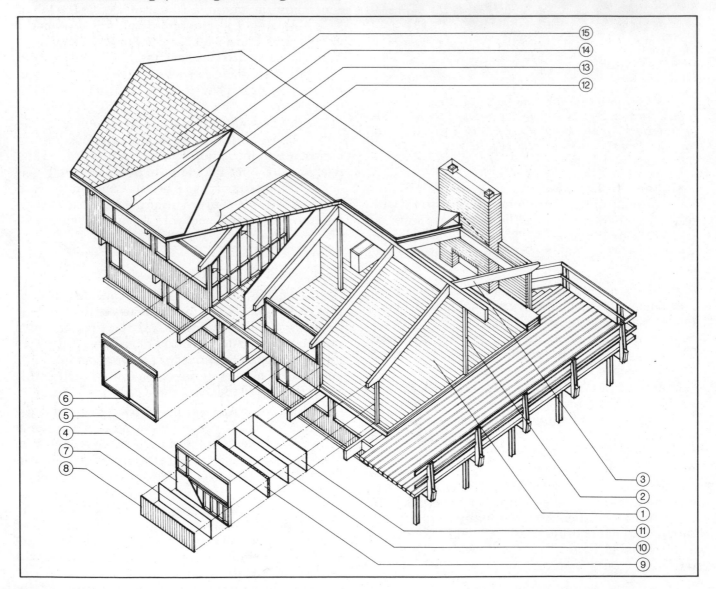

Elements of one precut, factory-built unit: Interlocking laminated decking (1) is nailed to the framework of posts (2) and beams (3) to form the floor and roof system. Exterior walls are installed next between the perimeter panels—consisting of (4) kiln-dried 2 x 4s and plywood sheathing, glass panels, framed (5) or sliding glass doors (6). Once these panels are in place, asphalt paper (7) and exterior siding (8) are applied. Wall insulation (9), a polyethylene vapor barrier (10), and an interior surface of Sheetrock (11) complete exterior walls. The exterior surface of the roof decking is covered with asphalt felt (12), a layer of rigid insulation (13), a second layer of asphalt felt (14) and asphalt roof shingles (15).

While erection cost of the basic structure may be low, that cost is just a fraction of the total cost. Total cost would include site preparation, shipping, plumbing, heating and air conditioning, wiring, and landscaping. The total cost, in fact, could range anywhere from three to nearly ten times the basic erection cost.

Solar Homes Solar systems for domestic hot water and space heating are the new frontier in residential energy conservation technology. Some packaged homes offer these features.

One manufacturer offers a Cape design with 47 degree roof pitch, on the belief that roof collectors should not monopolize the southern exposure and view. This solar system is designed to provide between 40 and 70 percent of the home's heating needs; a backup system such as an oil-fired warm-air furnace or wood-burning stove is needed.

These homes have been designed so that over-hangs shade the glass when the sun is high in the summer but let the sun's heat into the living space during winter. Sun-warmed air is distributed throughout the home by heating system return ducts.

Solar systems work best in homes that are properly insulated to cut down on heat loss through walls, roof, and floors. Standard construction of these homes includes double insulated windows, weather stripping, minimum windows on the north-

Here are two factory-built houses being constructed on site. The main structural elements have been constructed, and then side panels and roofing are added. Photos courtesy of Cluster Shed, Inc.

ern exposure, internal chimneys, maximum insulation in walls, roof, and floors, and entries designed to cut down on the amount of cold air coming into the home.

A packaged home with a solar heating system is expensive. Solar systems generally will add at least 10 percent to the cost because the technology is still being improved and refined.

Energy Saving Tips

Energy conservation is important everywhere. In vacation homes, it may or may not be of concern depending on the scope of the project, the number of seasons in use, and climate. Small cabins used only on a warm summer weekend really do not need many energy-saving features because very little energy is consumed. Large vacation homes used in all four seasons should have as many energy-saving features as primary homes.

Although it would take an entire volume to detail everything you need to know about energy conservation, in this chapter we will touch on the most crucial elements.

Insulation

Without question, the best barrier between you and high energy bills is insulation. It is the most important energy-saver you can install in the home. Even if you are just contemplating a weekend cabin, remember that if you decide to sell it sometime in the future, well-insulated walls, floors, and ceilings will increase value.

Insulation works most efficiently when it is applied to every area which is in contact with the exterior. This barrier keeps hot or cold air outside and reduces drafts. All insulation reduces the amount of heat transferred from a warm area to a cold area. When selecting insulation, the material with the highest R-value (resistance to heat flow) gives you the lowest transfer of heat. Today, other criteria must also be considered when selecting a specific kind of insulation.

There are many types of insulation. These include mineral fiber, cellulosic fiber, foamed plastics, and expanded mineral materials.

Mineral Fiber This is one of the oldest forms of modern insulation. Originally, it was manufactured by melting down slag, a by-product of steel production. Today, the product is either rock wool or fiberglass and comes in batts or blankets. It is an extremely effective type of insulation which can be used easily by the do-it-yourselfer in walls, crawl spaces, and ceilings.

The material is fireproof; however, the vapor barriers attached to the insulation are not. Normally it can be purchased with or without paper vapor barriers. Vapor barriers reduce the movement of moisture from a warm to a cold area and prevent moisture buildup in walls and attic areas.

Cellulosic Fiber This material is produced from wood pulp or another wood fiber material such as old newspapers. The material is not fireproof in itself, so a fire retardant material must be added. Some fire retardants can corrode piping and heating ducts.

Cellulosic fiber insulation produced by a legitimate manufacturer is a good product. When shortages of insulation occurred in recent years, some people started producing and selling this type of insulation without taking safety precautions. If you are purchasing this type of insulation, check that it is properly labeled with fire-retardant chemicals listed.

Foamed Plastics These include polyurethane, polystyrene, or urea formaldehyde. Often, this substance is used as packaging material for delicate items such as high fidelity components. Typically, it is molded around the product and holds it firmly in its box. Foamed plastics offer the benefit of a very high R-value. This insulation usually comes in boards and is easy to install. It must be used with a fireproof wall covering such as gypsum wallboard. Although this material is fire-resistant, when exposed to extremely high temperatures it can melt and produce a dangerous, toxic smoke.

Expanded Mineral Materials This insulation includes substances such as vermiculite and perlite. Expanded mineral materials are fireproof. Because of the nature of the material, it must be blown or poured into place. Many contractors are available with machinery to blow this insulation in place. If you want to insulate your own house, about the only place you can use this material is in the attic where you are able to pour bags of the granule material between ceiling joists.

How and Where to Insulate

If you are going to do the job yourself, we will assume that either you built the house yourself or you had a builder erect a rough shell. Rough plumbing and rough electrical should already be installed, and systems should have been tested before insulating begins.

Heat rises; therefore, the most crucial place to insulate is in the ceiling or attic. If you do not intend to use the attic, it is easiest to place the insulation between ceiling joists. If the attic is large enough and you plan to finish it off someday, the most practical place to insulate is between roof and rafters.

In either case there are a number of things you must do before insulating. Check to make sure you have adequate ventilation in the space. Mark the ventilators, and do not cover them with insulation. Next, locate all light fixtures. Do not insulate immediately around light fixtures. Whether you use loose fill or a blanket type insulation, build a simple wooden frame around the light fixture to keep insulation away from it. The sides of the box should be the same depth as the insulation you wish to install. Fiberglass insulation won't burn, but it is such a good insulator that it might cause the fixture to overheat.

Check the attic space to ensure that there are no leaks. Wet insulation is bad insulation.

Now consider a vapor barrier for your insulation. All vapor barriers face toward the warm part of the house. If you are installing insulation between ceiling joists, the vapor barrier goes down first. If you are insulating roof rafters, the vapor barrier faces in toward the attic space.

The most convenient insulation to install is batt or blanket insulation with a vapor barrier already attached. If, for instance, you have 8-inch ceiling joists in your attic, you can use either 6- or 8-inch insulation and roll it between joists with the vapor barrier facing down.

If you choose to use insulation without a paper vapor barrier, you will have to install a vapor barrier before you apply insulation. The best product to use in this case is plastic sheeting. To insulate between roof rafters, you will have to roll out the vapor barrier and insulation, and, as you do so, you will have to staple it to the roof rafter. Insulation should be stapled from the lowest point of the rafters on up to the peak of the roof. As mentioned, the vapor barrier should be visible when the job is done.

Attic sidewalls must also be insulated. Here, insulation is again rolled out and stapled in place.

Any time you handle insulation material, wear a mask or some other type of breathing apparatus. Tiny particles often break off and can be inhaled causing irritation and discomfort. You can protect your hands from this material by wearing work gloves.

For walls, use batts or blankets if you intend to do the job yourself. Again, purchase insulation with vapor barriers. Staple the insulation into place with the vapor barrier facing in toward the warm section of the house. Insulation should be stapled from the lowest part of the floor right up to the high point of the wall.

If you have oddly spaced sections of the wall which will not take a full blanket of insulation, rip off a portion and stuff it in place. All spaces around windows and doors should have insulation stuffed in before the final wall covering is added. Your aim is to seal off the interior of your house from the weather outside, and the only way to do that is to staple or stuff insulation in every exterior wall or space.

You may come upon plumbing and heating ducts in the wall. Where possible, insulate behind pipes and heating ducts to protect them from the exterior. They should never run through an uninsulated area. If you can not stuff insulation behind them, wrap pipes and ducts in special insulation that you can buy for the purpose. For pipes in particular, it is not just a matter of energy conservation but protecting the pipe itself from freezing and bursting.

Basements present another problem and another opportunity for energy saving, whether your house has a full foundation with masonry walls or a crawl space with masonry walls. You can frame out the full basement walls with 2 x 4 studs and staple insulation between wall cavities, or you can nail 1 x 2 furring strips to the walls with masonry nails and apply board insulation over the furring strips. The furring strips should be applied 24 inches on center if you are using 24-inch-wide boards. The walls must then be covered with wallboard.

Unless you plan on using the basement as living space, the easiest and least expensive way to insulate is to apply insulation in the cavities between floor joists. Stapling does not work well in this situation because gravity pulls insulation down after a while. An extremely effective way to insulate here is to apply the insulation and then staple a wire mesh to joists to hold it securely in place.

Many vacation homes are constructed on a crawl space. There might be either a crawl space constructed of masonry block or a crawl space created through pier construction. Insulation can be applied between floor joists in a similar fashion as in a full foundation. Another option is to lay insulation against the crawl space sidewalls. In either case, face the vapor barrier toward the warm part of the house. Lay a plastic sheet on the bare earth under the crawl space to help control moisture.

Windows and Doors

Windows Windows use up energy in several ways. Air often infiltrates around panes of glass and the window frame causing wasteful drafts. Glass is an excellent conductor of heat. It conducts heat out of the house in winter and into the house in summer. Windows also transfer heat through radiation.

Windows with wooden frames conduct less heat

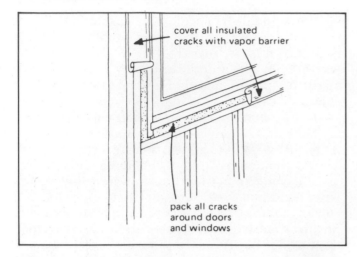

For a thorough insulation job, it is necessary not only to fill wall and floor cavities but also to fill small spaces such as those between the window frame and wall. Drawing courtesy of National Mineral Wool Insulation Assn.

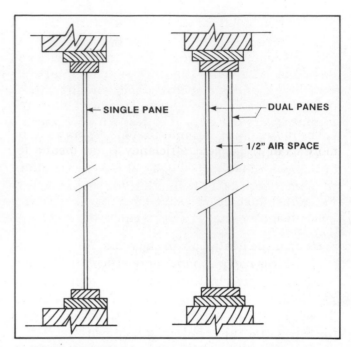

If you use double-glazed glass, you can usually reduce heat loss by about 50 percent. Drawing courtesy of Owens-Corning Fiberglas

than those with aluminum frames. The wood units, however, are more expensive. Before you decide which to buy, measure the potential cost savings against the initial cost of the windows.

Double-insulated glass is an extremely good energy saver if you plan to use the home during cold months. Big windows use up more energy than small ones.

Storm windows are necessary for regular winter use. If you choose not to spend the extra money on storms, you can still save energy by installing do-it-yourself storm windows made of clear plastic and held in place with strips of quarter-round molding.

Doors The real energy saver for doors and movable sections of windows is weather stripping. Weather stripping is made of either metal, plastic, or felt strips and is designed to seal between the movable sections of windows and doors and thus cut down on energy waste and cold drafts. Metal weather stripping holds up best and is most expensive.

There are a variety of insulated doors on the market. They usually have metal faces with insulation sandwiched in between. The ordinary wood door offers some insulation but not nearly as much as a well-insulated wall. Storm doors will make a wood door more efficient.

Caulking Caulking is a material which insulates cracks in the siding and between window and door frames. Caulking is necessary not only to keep out cold drafts, but also moisture.

Caulking comes in disposable cartridges and is applied with a caulking gun. Three types are available, including: oil or resin base; latex, butyl or polyvinyl base; and elastomeric.

Oil- or resin-base caulking will bond to almost

any surface. It is not very durable, but it is cheapest. The latex, butyl, or polyvinyl caulking will bond to most surfaces. It lasts longer and is more expensive. The elastomeric caulkings are the most durable and the most expensive.

Heating and Cooling Systems

Depending on the scope of your vacation home, you may or may not install a central heating system. These systems are expensive, and unless you use your house frequently in cold weather, they are unnecessary.

If you need one, carefully check the price and availability of fuel in the vacation home area. Although natural gas may be available and relatively inexpensive, extending a supply to your house may be costly. In many areas bottled gas is available, but this is a very expensive way to heat a home.

Oil is another possibility, but make sure that an oil delivery truck can negotiate your road. This will require another excavation on your building site for the oil tank.

Electricity is very expensive, but it might be necessary if you need limited winter heating. It is easy and relatively inexpensive to install, but if used extensively over the years, the cost savings will disappear.

If you use your house infrequently during winter, good electric resistance heaters may be desirable. While most people interested in energy conservation would balk at using such devices, they will heat a cold house.

Fireplace Heat For many vacation home users, a central heating system is not needed because the house is only used in spring, summer, and autumn. Basically, these people need something to take the chill out of the air in the early morning and evening. In most climates, an efficient fireplace could achieve this.

The typical conventional fireplace operates at no more than 15 percent efficiency. That means for every log you burn, you only derive about 15 percent of the potential heating power. Besides having low efficiency, most fireplaces have an extremely efficient flue, which is not totally desirable. An effective flue will not only remove smoke, but a good portion of the heated air in the room.

Fireplaces can be made more efficient. The flue, for instance, should always be closed when the unit is not in use. A good airtight cover will keep much of the air in the room while the fireplace is in operation. There are grates on the market which enable more heat to move into the living space.

When constructing a masonry fireplace in a new vacation home, the main section of the chimney should run up through the interior of the house rather than up an exterior wall. Masonry walls heat up while the fire is burning. As the house cools down in the evening, the masonry walls will give up heat to help warm the house.

Vacation homeowners generally can not afford to spend as much on their second home as they do on their first. Therefore, a freestanding metal fireplace may be more economical. Many of the same energy-saving techniques apply to the freestanding unit.

A properly ventilated attic will greatly reduce temperatures in the living space below. Drawing courtesy of Leslie-Locke

Freestanding fireplaces can be installed by a do-it-yourselfer by following manufacturer's instructions. The only difficult part is cutting a hole in the roof. You can hire a carpenter to cut the hole.

Wood-Burning Stoves As a source of extra heat, a quality wood-burning stove is a good choice. Although they are not as beautiful as fireplaces,

There are a number of heat-circulating fireplaces on the market today that can greatly increase efficiency. This unit is installed by (1) cutting a hole in ceiling and roof; (2) installing insulated pipe; (3) framing out metal unit with 2 x 4 studs, and (4) covering framing with plywood and a covering such as paneling or simulated brick or stone. Photo courtesy of Heatilator

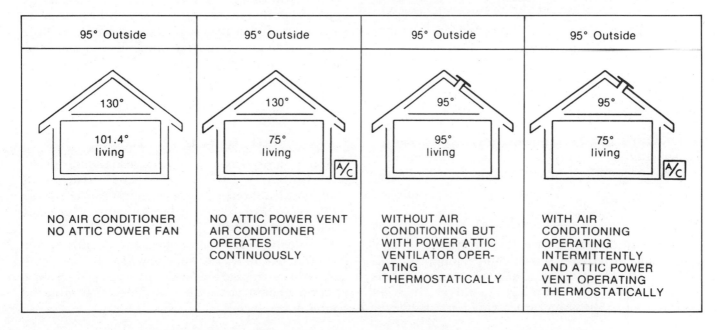

95° Outside	95° Outside	95° Outside	95° Outside
130° / 101.4° living	130° / 75° living	95° / 95° living	95° / 75° living
NO AIR CONDITIONER NO ATTIC POWER FAN	NO ATTIC POWER VENT AIR CONDITIONER OPERATES CONTINUOUSLY	WITHOUT AIR CONDITIONING BUT WITH POWER ATTIC VENTILATOR OPERATING THERMOSTATICALLY	WITH AIR CONDITIONING OPERATING INTERMITTENTLY AND ATTIC POWER VENT OPERATING THERMOSTATICALLY

stoves are more energy efficient. A good wood-burning stove is more efficient because it draws just enough air for combustion and little more. More of the heated air remains in the room. As the wood burns, the metal unit heats up and, by conduction and convection, gives off considerable heat.

There are traditional wood-burning stoves called Franklin stoves that are efficient and beautiful. A good unit is expensive, but will last for many years. There are cheaper models available, but the heat source is one item in your vacation home that you should not skimp on.

Whether you buy a new or used unit, look closely at all seams. A welded seam that is rough, pocked, or cracked indicates that the unit is constructed poorly or damaged. A crack will only grow larger.

Purchase a wood-burning stove from a reputable dealer. When buying a unit, be sure delivery is included in the price. A good stove weighs several hundred pounds and is difficult to handle. Some dealers will also install the unit for you for a fee. If you decide to do it yourself, make sure the unit has detailed instructions.

Another option is to build a wood-burning stove. There are kits available to make a wood-burning stove from a 55-gallon drum. This type of stove is not as efficient as a Franklin-type stove, but it is a good, inexpensive source of heat. Before converting a steel drum thoroughly clean out all combustible fluid before cutting.

Water Heaters Central heating may not be desired, but you will want a hot-water heater. You may want a standard hot-water heater, or you can install a solar hot-water heater. Although solar heating systems are not very practical for a vacation home, a solar hot-water heater can be.

Standard hot-water heaters have between a 30- and 50-gallon tank. A 50-gallon tank is most common in first homes. If hot water usage will be less at your vacation home than at your first home, select a smaller size.

Remember to shut the unit off between visits to the house. Keeping your hot-water heater on between visits will boost your bills to wasteful proportions.

Insulating hot-water pipes always saves energy. This is quickly accomplished by wrapping insulating tape around hot-water pipes.

Ventilation

Good ventilation, particularly in the attic, contributes to lower energy bills and a more comfortable environment. Even if energy conservation is not a major concern, comfort should be. On a hot summer day the temperature could reach 130 degrees in an unventilated attic. At night, as it cools down, that trapped heat radiates into the living space and keeps everything uncomfortably warm. If hot air can be trapped in the attic, so can water vapor. If moisture remains in the area over a long period of time, dry rot can develop.

The solution is simple; provide ventilation. A power ventilator is expensive, but it changes the stale air quickly and efficiently.

Usually, a combination of static vents is best. Static ventilators are adequate for most climates. These units are built into the structure and replace hot or stale air slowly but efficiently.

Roof Louvers These are small aluminum, steel, or plastic domes located near the ridge of the roof. They are available with or without screens. Units with screens block airflow slightly, but they do keep insects out of the home.

Turbine Wheel This is a type of roof louver. The unit comes with a wheel or turbine which turns in the wind and draws air out of the attic. In severe weather, rain enters through both the turbine wheel and the roof louver.

Gable-End Louvers These simple units are placed at the ends of the house and work most efficiently when the wind blows through them.

Ridge Vents This is a long vent which provides continuous circulation of air along the ridge of the house.

Soffit Vents This vent offers airflow along the floor of the attic.

The best combination of vents is a ridge vent in conjunction with a soffit vent. This positions vents at the highest and lowest points of the space.

Ventilation in other areas of the house should be considered during design. A simple way to achieve good ventilation is to have windows and doors on opposite walls to allow cross-ventilation. Also consider a small fan in the kitchen and bathroom to help remove moisture from the house. A small exhaust fan is recommended.

Good Maintenance

A key energy-saver in any home is regular maintenance. Empty your hot-water heater to drain the sludge at least once a year. Foreign matter in the water collects in the hot-water tank and reduces the efficiency of the unit.

Maintenance of water faucets should also be routine. A dripping faucet wastes water and money.

Weather stripping and caulking should be checked at least once a year. Damaged or worn material should be replaced.

Space Savers and Maintenance

Planning a Small Kitchen

A kitchen is more than just a place to prepare and cook food. In vacation homes the kitchen also serves as a center of activity and entertainment. To create a pleasant but efficient small kitchen, the key elements are: appliance layout, cabinets and other storage space, countertop work space, lighting, and dining areas or counters.

Kitchen appliance manufacturers and cabinet designers have conducted considerable research to discover the most efficient appliance layouts. Food preparation, cooking, and cleaning tasks take place within a work triangle. This triangle can be roughly drawn from the refrigerator to the rangetop to the sink. If the sides of the imaginary triangle are equal, the kitchen will be efficient. For maximum efficiency, the sides of that triangle should total a minimum of 16 feet.

For an efficient layout of appliances, cabinets, and countertops, there are L- and U-shaped kitchens. Single-bowl sinks take up less space than double-bowl sinks and cost less.

Wall cabinets, pantries, and base cabinets come in a standard 12-inch depth that provides ample

In this L-shaped kitchen area, natural light has been used to its fullest. Photo courtesy of California Redwood Association

storage space. Analyze your family needs and plan storage space accordingly.

Storage space in standard cabinets can be increased with ready-made hardware: lazy susans, plastic dish and utensil racks, adjustable shelves made of a variety of materials including plastic bins, vinyl-covered steel rods, and plywood.

Use the space above wall cabinets for storage. Space under wall cabinets can be used for narrow shelves to hold spices or teas. Hang pots on the wall, or hang them from a wrought iron or wood ceiling fixture.

Plan to place a window in the center of an exterior wall, preferably over the sink. This will let in natural light, especially if it faces south or east.

You can build an open kitchen, separated from a dining/living area by an island of cabinets and countertop. For informal dining you can sit on stools and eat at the countertop. In smaller cabins, kitchen appliances and cabinets can be located along one wall with no physical separation of kitchen and dining areas.

Multipurpose Spaces

Since interior space is limited, be sure to use it efficiently. Avoid poor traffic patterns and space wasters like hallways. Plan on using interior space for more than one purpose.

Multipurpose rooms maximize space in vacation homes. The room shown here is used as a den, entertainment center, or guest room. A pull-down bed hides in the wall, covered by plywood paneling. Photo courtesy of Champion Building Products

The interior of this vacation home is kept simple with the kitchen installed along one wall and an open area for dining and seating. Window seating is one way of maximizing interior space use. Photo courtesy of McCue, Boone, Tomsick Architects

Bedrooms can be used as hobby rooms, dens, or sitting rooms. Other living areas can provide additional sleeping space. Place a bed lengthwise against the wall.

The beauty of a vacation home is that you can create dramatic and functional effects that may not be suitable for your primary home. You can create extra space for sleeping, reading, or conversation by building a loft. A simple wooden ladder will suffice in making the loft space accessible. Knocked-down preassembled staircases made of wood or metal are also available and are cheaper than custom-built units or standard staircases. Special attention should be paid to the ventilating, heating, and cooling of loft areas. Heat rises, and without proper ventilation these spaces can become intolerably warm. Protective railing is needed, particularly if you have children.

Planning Storage Space

Storage space is a valuable commodity in any home. Both built-in and freestanding storage space should be carefully planned to bring order into your vacation world.

Since you probably will be bringing in bulky sports equipment like skis, fishing poles, or golf clubs and tracking in dirt at the same time, a mudroom would be a useful addition to your house plan. Seating could be built around the perimeter of the room to allow people to sit and comfortably remove wet things like flippers or ski boots before entering the main part of the home. Hooks and shelving in this room provide an orderly space in which to store boots, shoes, equipment, and outdoor clothing.

Several attractive possibilities are available to maximize storage in the main living area. You can use wall space under windows for seating and storage units.

Built-in wall units accomplish two purposes as interior decor and as handy places to store clothing, books, towels, sheets, and other necessities. You can easily build the units yourself. There are many free-standing wall unit systems you can buy. They come in several styles, and components can be mixed to give your wall a custom look.

Wall units come in materials ranging from Formica over particleboard or plywood to hardwood; costs vary too, from a few hundred dollars to several thousand dollars. Some freestanding systems come with "cap" units that fit over top components; standard height is 78 inches, and the "cap" units will bring the wall system up to a standard height of 8 feet.

Look for unexpected spaces in your vacation home that might provide storage. The wall under a staircase can be used for storage. Don't overlook wall space above doors. In A-frame designs, storage cabinets can be built into the wall and under the roof on the second floor loft area. If you make your own furniture, consider making couches with a hidden storage area beneath the seat.

Platforms—The Ultimate Built-In

A platform system in the living area can provide storage, seating, and even sleeping space. If your tastes are modern, a multilevel platform design can even serve as chairs, sofas, or end tables.

You can build a plywood platform 8 to 12 inches high into your living-area design along one wall of the room. For storage, add drawers along the perimeter of the platform. The interior part can be made accessible with lift-up covers. Platforms can be covered with an industrial grade carpeting which wears well. Add some pillows and the platform can serve as functional seating or sleeping space.

More Storage Space

The space in between interior wall studs can be used for storage. Plan for this type of space before construction, rather than after the house has been finished. For the kitchen and bathroom, recessed cabinets with adjustable shelving can be installed into the space between the studs.

Room Dividers If the open living-area design is not to your liking, you can create a room divider which also functions as a storage unit. Several room divider plans, suitable for living/dining areas or living areas/sleeping alcoves, are available from

Kitchen/dining areas can be separated by a room divider that also acts as buffet-server or countertop for informal dining. A window over the sink and range brings the outdoors inside. Photo courtesy of Champion Building Products

the American Plywood Association, 119 A Street, Tacoma, Washington 98401.

In the bedrooms, efficient storage starts with the closets. Here are a few tips to create ample storage space in your closets:

- Make walk-in closets wide enough for storage on both sides.
- Install hooks, racks, or shelves on the back of closet doors.
- Make shelves adjustable.
- Use double rods where possible, to increase hanging space.
- Use ready-made organizers, such as stacks of plastic boxes or wire racks for shoes and multi-purpose garment hangers.

Wall units are suitable for bedrooms, too. They can replace standard dressers and armoires, which take up more space. Place them opposite the bed, or work them into the wall space around the bed.

Some platform beds come with built-in storage. Unpainted furniture stores are a good source for this type of bed.

Lighting

The home is not complete unless provisions are made for interior lighting. Where intensive lighting is needed, as in the kitchen, use fluorescent fixtures. They use less energy and last far longer.

Extra sleeping space created in and below loft area. Lofts are most appropriate when ceilings vault to a height of 20 or 25 feet.

Indoor/outdoor living suits vacation retreats. Deck area off this living room could be used for dining in warm weather. Photo courtesy of Champion Building Products

In living areas and bedrooms, wall-hung exposed white bulbs can be used effectively. They give off a great deal of light, with little glare. Dimmer controls will extend their life and reduce electricity consumption. Low-cost "industrial" style hanging fixtures are good for accent lighting in conversation and dining areas.

You can reduce the number of lighting fixtures needed in your home by taking advantage of natural light. The amount of light coming in depends on the amount of sky visible through the window. The higher the window area, the more natural illumination. Insulated domed skylights diffuse natural light over a broad area. They also reduce heat transmission and, therefore, lower heat gain during the summer. In the winter when sun angles are lower, they have better light transmission than flat skylights. Skylights can be effectively positioned in living/dining areas, kitchens, or bathrooms.

Materials for Minimum Maintenance

The interior of a vacation home should be designed for minimum upkeep. Materials in a vacation home that require constant attention and cleaning or that are susceptible to deterioration will take away from valuable vacation time.

Unfinished wood such as cedar, knotty pine, or redwood can be used as interior finishes to create a warm atmosphere that will also provide years of low-maintenance use.

Choose a floor covering that is durable and easy to clean, such as vinyl asbestos tile (particularly for high-use areas). You can cover floors with area rugs for an attractive appearance.

Using Formica for kitchen countertops is economical and easy to maintain. Open shelving in the kitchen adds low-cost storage space but must be cleaned more frequently than standard cabinets.

Protecting Your Vacation Home

Protecting your vacation home is harder than protecting your primary residence because it is unoccupied for longer periods of time. Vacation homes are generally located in a secluded area where it is hard for local authorities to keep a close watch. Protection of the dwelling should begin with the first shipment of materials to the building site.

The Building Site

Even in distant, serene surroundings, thefts can take place. If you leave building materials and tools unprotected at your vacation homesite, you are inviting trouble. The best way to avoid problems is to leave nothing at the site that has not been nailed down. If you are building the house yourself, only take delivery of materials which you will use at that particular time. This is not always possible, but you should try.

Concrete block can usually be left in the open because a thief would have to labor long and hard to steal them, and replacement cost would not be that great. Never leave a mechanical concrete mixer unprotected. This is an expensive item to replace. Chain it and lock it around a thick tree on the building site.

If you are building a home on a concrete block crawl space or full foundation, get the block in place as quickly as possible, get the bottom platform in place, and put a door on the foundation area. This can be excellent space to store materials although it is not 100 percent effective in preventing losses. This secured area is a good place for wheelbarrows, shovels, and other such items but not for easily removed and expensive tools. Keep those in the trunk of your car, and take them to and from the site as you need to use them.

If a contractor is building your house, he is responsible for all materials and his tools. It is best to have a contract with him making him responsible for materials and labor.

If you are building the house yourself, you will have a lumber list so you can estimate costs. Do not take delivery of all lumber at once. To get the best deal on costs, negotiate a package price with the lumberyard, but have them make deliveries in small quantities. If a theft occurs you lose only a portion of your lumber. If you are a weekend builder, store as many of the materials as you can.

Once the superstructure has been completed, put all windows and doors in place and lock them securely. Besides thieves, building sites are also subject to vandals. Leave nothing around the building site that a vandal could use to do damage.

Have adequate insurance for your home from the time you begin the project. Usually you need a contractor's liability policy while construction is underway, which then converts to a homeowner's policy upon completion.

You will need electricity at the site at the start of the construction process in order to operate power tools. Lights with timers can deter vandals and thieves once the superstructure has been completed. Notify local authorities that the house is being constructed. Although the police can not keep constant watch on the site, they may cooperate and drive past it occasionally.

As construction proceeds, do not leave expensive, uninstalled windows, cabinets, or appliances in the house for any longer than necessary.

Protecting the Finished House

There is no way to assure absolute protection of your vacation home or belongings while you are not there. Although windows and doors can be locked, they can easily be broken. Use well-made security hardware that will not only protect the home while you are not there, but also while you are there.

When you are not going to be at the house for long periods of time, it is best to remove valuables, whether an antique quilt or rocking chair or your favorite painting. If the house is broken into and robbed, you will want to minimize your loss as much as possible. Common sense should be exercised.

Animals If your new house is located in a wild area, you will have to protect it from animals. If there is nothing in your cabin worth stealing, you should still secure it. A raccoon can cause a big mess. For the same reasons you will also want to keep any trash under lock and key. A latch can be opened by a raccoon or other animal.

Smaller animals such as rodents and termites can also damage your vacation home. Rodents generally have two ways of entering the house: underneath a foundation or through the flues. Concrete foundations are a better deterrent than the other types. Flues should be capped tightly if you are closing the home for an extended period of time. Materials need not be fancy; a metal hood or wire screens will do. If your stovepipe is not hooded, that should be capped. Dampers should also be closed.

Do not give animals any incentive to enter your home. Remove opened food and liquids because they will attract animals and insects. As a safeguard against rodents, store bedding inside bunk frames or storage boxes lined with metal.

Termites Termites feast on the wood in your house, preferably wood that is decaying in an environment of warm temperature and high soil moisture. Subterranean termites, responsible for most termite damage in the U.S., build nests and subway systems well below frost line, near their food supply—the house.

Their tunnels are continuous and clear cut. If the above-ground path to the wood is interrupted by an impermeable surface, they will build a small mud tube to the wood. Once they reach their goal, your flooring, studding, window and door frames are all vulnerable.

Consider these measures to dissuade termites from becoming occupants.

- Treat soil with a poison at the building site; a professional should do this.
- Use preservative-treated wood for parts of the superstructure subject to termite invasion.

- Use metal casings for basement windows and doors.
- Remove all dead wood and stumps from areas of the building site near the house—don't just bury them.
- If you have a crawl space instead of a full foundation, install an access window and ventilation. Closed-off areas under the house should be avoided.

How do you recognize a termite problem? Unfortunately, in many cases you don't recognize the problem until damage is done. Look for termite tunnels in the basement or crawl space. If tunnels are moist, a colony is in residence below your home. If tunnels are dry and brittle, the termites may have moved elsewhere. Keep checking for cracks in the foundation where termites can enter. The termites that make their forays into the wood to bring food back to the colony can penetrate a masonry crack $1/64$ inch wide. If you can push the point of a small knife or screwdriver about $1/2$ inch into any wood in the basement, it may be a sign of trouble.

If your home suffers termite infestation, call a professional exterminator to chemically treat the house. Get a one-year guarantee. Annual inspections are a good idea.

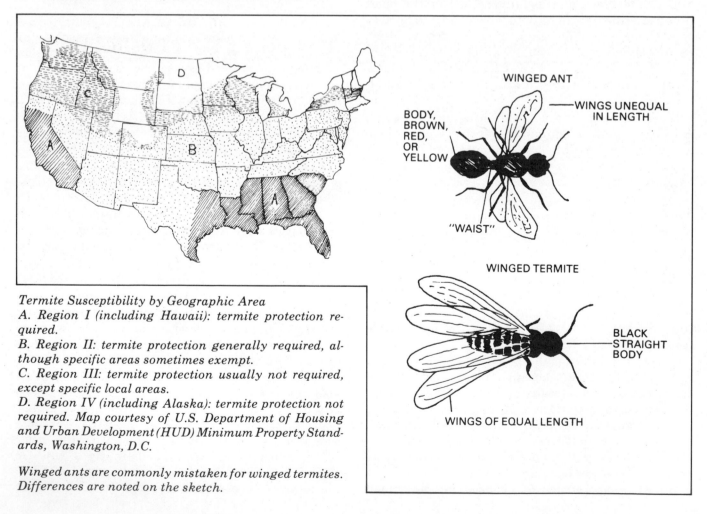

Termite Susceptibility by Geographic Area
A. Region I (including Hawaii): termite protection required.
B. Region II: termite protection generally required, although specific areas sometimes exempt.
C. Region III: termite protection usually not required, except specific local areas.
D. Region IV (including Alaska): termite protection not required. Map courtesy of U.S. Department of Housing and Urban Development (HUD) Minimum Property Standards, Washington, D.C.

Winged ants are commonly mistaken for winged termites. Differences are noted on the sketch.

WINGED ANT

BODY, BROWN, RED, OR YELLOW

WINGS UNEQUAL IN LENGTH

"WAIST"

WINGED TERMITE

BLACK STRAIGHT BODY

WINGS OF EQUAL LENGTH

Fungi Fungi are microscopic plants that can devastate wood. Some fungi only discolor wood, while decay fungi destroy fiber. Fungi can not work in dry wood; they need moisture. Fungi and soil-nesting termites can work in the same wood.

To prevent decay, dry wood should be used in the construction and kept dry during the building process as much as possible.

The sapwood of all tree species is susceptible to decay, whereas heartwood is more durable. Woods treated with a preservative are more decay-resistant.

Roof design can help prevent decay. An overhang of 12 inches on a one-story house is effective. If there are no overhanging eaves, gutters and downspouts are desirable. Wood surfaces like windows and doors should be flashed with a noncorroding metal.

Wood should not be allowed to come in contact with soil unless the wood is thoroughly impregnated with the right preservative. Wooden items, such as grade stakes, concrete forms, or stumps left on or in the soil under the home, also invite decay.

Other factors that contribute to decay are:

- Undrained soil and insufficient ventilation in homes that have no basements.
- Wood parts embedded in masonry near the ground.
- Use of infected lumber.
- Poor joinery around windows and doors and at corners and inadequate paint maintenance.
- Attics that are unventilated.
- Roof leaks, and leaks around kitchen and laundry equipment and shower/bathtubs.

Proper maintenance must be practiced regularly. Watch for such things as leaky roofs, clogged drains, rust, and sweaty pipes. If fungi are discovered, they should be traced to their source of moisture, and the situation corrected. Properly cured wood should be used to replace wood that has been hopelessly decayed. If the infection has spread, wood 2 feet in each direction from the decayed part should also be removed and replaced. Before putting the new wood in place, all old wood and masonry surfaces around the decayed part should be brushed with a preservative.

Protection from Fire

A fire can be devastating to your home, whether or not you are in it at the time. While you are in the house, be careful with matches and chemicals, particularly if there are children around. Do not smoke in bed; cigarettes and matches are the major causes of home fires.

Be sure that electric wiring is safe and adequate.

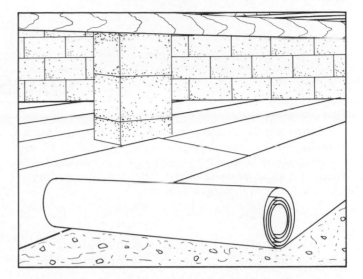

To keep soil moisture from vaporizing into the air and then condensing on joists and sills, cover soil under house with roll roofing. This technique also enables you to close crawl-space ventilators in winter. *Drawing courtesy of U.S. Dept. of Agriculture*

Electric circuits should be fused properly. Lightning rods should be properly installed and grounded as should arrestors on radio and television antennas. Repair defective chimneys, spark arrestors, flues, stovepipes, and heating and cooking equipment if necessary. If you must store gasoline or other flammables, keep them in proper containers and locations.

Protection from Flooding

Flooding from severe storms or quick thaws represents a serious threat to many vacation homes. If flooding is common in your area, consider stocking plastic sheeting, sandbags, and lumber with which to build dikes around the house. Buy check valves from hardware or plumbing supply stores to prevent water from backing up through sewer traps and drainpipes.

If flooding is forecast, get objects off ground level by at least 2 or 3 feet. Keep underground fuel tanks filled and seal them so that fuel oil will not spread over the property in case of a flood.

There is an insurance program called the National Flood Insurance Program with coverage by the Federal Housing and Urban Development Department. Not all communities qualify, but you should check with a local property and casualty insurance agent. When a community first qualifies under the program, it is in emergency status. Once HUD sets premium rates, emergency status ends. You can then get considerably more coverage on the house as well as on the contents.

Glossary

Anchor bolts Bolts which secure the wood sill plate to the foundation wall.

Attic ventilators Either static or mechanical devices to move air in and out of the attic. Vents are needed to prevent heat and moisture buildup.

Backfill Replacement of earth against house foundation.

Batten Strips of wood used to cover joints or as decorative vertical members over plywood or wide boards.

Batter boards Horizontal strips of wood nailed to stakes set at the corners of an excavation. These boards are used to indicate the desired level. When strings are attached, they also help outline the foundation walls.

Beam Structural member transversely supporting a load.

Bearing partition A partition in a house or other structure which supports any vertical load and its own weight.

Brace A piece of lumber which, when applied to a wall or floor, will help stiffen the structure. Temporary bracing is used in home construction to keep walls or other members steady and level while work proceeds.

Bridging Wood or metal members which are used in a perpendicular or diagonal position between joists at midspan to act both as tension and compression members to brace joists and spread the action of loads.

Butt joints A joint where the ends of two pieces of lumber or other material meet in a square-cut joint.

Collar beam A wooden member, usually one or two inches thick, which is used to connect opposite roof rafters. This beam helps stiffen the roof rafters and the entire structure.

Column A vertical member used to support loads.

Condensation Droplets of water which condense from water vapor. This usually occurs in a house in an unventilated attic or basement. If the condition persists, it can cause dry rot on wooden members.

Corner braces Diagonal braces which are applied at corners to help strengthen the wall.

Cornice Overhang of pitched roof at the eave line.

Crawl space Shallow space below a house, usually enclosed with foundation walls.

Crowning Term used to identify which way joists should be installed. Virtually every joist, on end, will have a slight warp. The side which warps up should be placed up. When plywood subflooring is applied, the joists straighten out.

Dormer An opening in the slope of the roof. The framing projects out and the dormer end is suitable for a window. In small homes where there isn't much of a second floor, the addition of a single or double dormer can add to the living space of the house.

Downspout Pipe for carrying water from roof gutters.

Fascia A horizontal board which is used as facing.

Footing Base for foundation walls, posts, or chimneys which is usually made of concrete. The footing is wider than the structure it supports in order to better distribute the weight over the ground to help prevent settling.

Gable The triangular portion of the end wall of a house with a pitched roof.

Gussett Small piece of wood or metal which is attached to corners or intersections of a frame to add stiffness and strength.

Header Framing lumber used around openings to support free ends of floor joists, studs, or rafters.

Header joist Horizontal lumber that butts against the ends of floor joists around the outside of the house to add strength and to tie joists together.

Joists Parallel framing members used to support floor or ceiling loads.

Molding Wood strip, either rectangular or curved, used for decorative purposes.

Nonload-bearing wall An interior wall which does not support any weight other than its own.

O.C. Abbreviation for on center. Joists, rafters, and wall studs are always positioned on center.

Outrigger Extension of rafter beyond wall line.

Panel Thin, flat piece of material used in house construction, such as plywood and wallboard panels.

Paper, building General term for all felts, papers, and other materials used in house construction.

Partition A wall which divides a space. An interior partition in a house generally divides a space into two rooms.

Pier A column, usually masonry but it could be of other materials, to help support other structural members.

Pitch Indicates the incline, or slope, of a roof. The pitch of a roof is expressed in inches per foot.

Plate General term used in construction. The sill plate is a horizontal member which is anchored by bolts, usually to the foundation. In wall construction, there is one bottom plate and two top plates. The plates here are usually 2 x 4 on the horizontal which are nailed into wall studs.

Plumb The exact perpendicular, or vertical. When a wall is plumb, it is exactly vertical.

Ply Denotes the number of layers of a material.

Plywood A piece of wood made with an odd number of layers of veneer. By gluing the wood veneers together, a strong material is created which far surpasses the natural material in strength.

Preservative Any substance which, when applied to a material, prevents that material from rotting or decaying.

Quarter round Molding which has the cross section of a quarter circle.

Rafter A structural member used to support the roof. When rafters are used on a flat roof, they are often called roof joists.

Rafter hip A structural member supporting the roof which forms the intersection of an external roof angle.

Reflective insulation A sheet material such as aluminum foil having one or both sides with comparatively low heat conductivity.

Ribbon A board horizontally on studs to support a ceiling or second-floor joists.

Ridge The horizontal line at the top edge of a roof where two slopes meet.

Ridgeboard Board placed at the top of the ridge to which roof rafters are attached.

Rise The vertical height of a step or flight of stairs.

Riser Vertical boards closing the spaces between the treads of stairs.

Roof sheathing Sheets of plywood or another material which are applied to roof rafters to enclose house. Roofing material is applied to roof sheathing, usually after an underlayment is added.

Sash A single, light frame containing one or more panes of glass.

Saturated felt Felt impregnated with tar or asphalt which is used in construction.

Shake A handsplit wood shingle.

Sheathing Structural covering used over rafters or studs. The material is usually lumber or plywood. Generally, walls and roofs today are sheathed with plywood panels.

Shingles Roof covering, usually made of slate, wood, asbestos, or a variety of other materials.

Shingles, siding An exterior wall covering, generally made of wood shingles or shakes or other nonwood material.

Siding On exterior walls, it is the finish covering. Siding can be of any material, from plywood to lumber to nonwood materials.

Sill The lowest member of the frame of a structure which generally sits on the concrete block foundation, usually known as the sill plates.

Sleeper Material, usually wood embedded in concrete, to serve as a fastener and/or to support subfloor or flooring.

Span The distance between structural supports, such as walls or columns.

Square As in a square of shingles. In construction, it is a measurement of 100 square feet.

Stud A vertical wall support member, usually wood or metal. In frame construction, a series of 2 x 4 studs is the vertical support attached to one 2 x 4 bottom plate and two 2 x 4 top plates.

Termite shield A metal sheet which prevents the passage of termites into the house. The material is usually put around the foundation and pipes near ground level.

Trim Finish materials in a house, such as moldings.

Truss A frame designed to act as a beam over long spans.

Underlayment Any material placed under finished coverings, such as shingles or flooring which provides a good surface for the finish.

Vapor barrier Any material which is used to retard the movement of moisture, mainly used in conjunction with insulation.

Weather stripping Narrow sections of metal used around doors or windows to prevent air infiltration.

Index

Block, concrete, 28-30

Caulking, 84
Ceiling, framing, 52-53
Climate, 4, 14
   beach, 14-15
   desert, 15
   heat, 15
   ski country, 15
   snow, 4, 15
   wind, 4, 15
Contracting
   architects, 18
   construction schedule, 13
   do-it-yourself, 10, 25
   finding builders, 10
   payment schedules, 10-11
   working with subs, 10-11

Decks, porches, 75-76
Doors, 74-75, 83-84

Financing, 7-8
Fireplaces, 85
Floor joists, 32-33
Flooring, 58
Footings, 28
Foundation, 28-31
   block, 29-31
   footings, 28-30
   layout, 28-29
   posts, girders, 30-32

Geodesic domes, 17, 80-81
Girders, 30-32

Heating, cooling systems, 84-86
Home types and plans, 10-11
   barns, 16
   cabins, 16, 59-77
   geodesic domes, 17, 80-81
   packages, 18, 78-79
   stock plans, 18

Insulation, 82-83

Kits, prefabs, packages, 78-81

Lighting, 89-90
Log cabin, 59-77
   assembly, 63
   building from scratch, 66-74
   interior, 68
   joists, 64
   materials, 63
   roofing, 65-66
   supplier, selecting, 60-63
   tools, 62-63

Maintenance, for
   energy savings, 86
Mobile homes, 80
Modular homes, 78-79

Piers, 23
Pole buildings, 23-24
Posts and girders, 30-32
Prefabricated homes, 80

Rafters, roof, 53-54
Roads, 6
   building, 19

Security
   during construction, 4, 91
   insurance, homeowners,
   flood, 91, 93
   protection against fire, 93
   protection against flooding, 93
   protection against fungi, 93
   protection against termites,
   animals, 92
Sewage handling, 6, 20-21
   absorption, 20-21
   septic systems, 21-22

soil percolation tests, 20-21
Sheathing, roof, 55
Shingles, 55-56
Siding, 51-52
Site selection, 5-6
   building site, 22
Space, guidelines, 14
   closets, 89
   economic design, 15-16
   kitchen planning, 87
   lighting, 89-90
   multi-use spaces, 87-88
   room dividers, 88-89
   storage, 88-89
Stairs, 76-77
Stoves, wood-burning, 85-86
Subfloor, 49-50

Taxes, real estate, 6
   tax consequences of renting, 8-9

Underlayment, roof, 54
Utilities, bringing in, 6, 19
   electricity, 19

Ventilation, 16, 86

Wall
   covering, 57-58
   framing, 50-59
Water, 6
   floodplains, 21
   moisture damage, 21
   storm damage, 5-6
   wells, 19-20
Water heaters, 86
Windows, 74, 83-84
Wood, properties, 59

Zoning, 6, 24
   building departments,
   permits, inspections, 24